Charles Slagg

Water Engineering

A Practical Treatise on the Measurement, Storage, Conveyance and Utilisation of

Water for the Supply of Towns, for Mill Power and for Other Purposes

Charles Slagg

Water Engineering

A Practical Treatise on the Measurement, Storage, Conveyance and Utilisation of Water for the Supply of Towns, for Mill Power and for Other Purposes

ISBN/EAN: 9783337139315

Printed in Europe, USA, Canada, Australia, Japan

Cover: Foto ©berggeist007 / pixelio.de

More available books at **www.hansebooks.com**

WATER ENGINEERING

A PRACTICAL TREATISE

ON THE

MEASUREMENT, STORAGE, CONVEYANCE, AND
UTILISATION OF WATER FOR THE SUPPLY
OF TOWNS, FOR MILL POWER, AND FOR
OTHER PURPOSES.

By CHARLES SLAGG,
WATER AND DRAINAGE ENGINEER, ASSOCIATE MEMBER OF THE INSTITUTION OF CIVIL ENGINEERS,
AUTHOR OF "SANITARY WORK IN THE SMALLER TOWNS, AND IN VILLAGES."

LONDON
CROSBY LOCKWOOD AND SON
7, STATIONERS' HALL COURT, LUDGATE HILL
1888

LONDON:
PRINTED BY WILLIAM CLOWES AND SONS, LIMITED,
STAMFORD STREET AND CHARING CROSS.

PREFACE.

SINCE a series of articles on 'The Water Question,' by the Author, appeared in the 'Building News,' frequent requests have been made that they should be reprinted in book-form. Those requests have now been complied with, and the Author has, at the same time, added much new matter, and rewritten some parts of the old.

So large a subject could not be treated from the Author's own experience alone, and he has freely referred to such sources of information as were available to him, for those parts of the subject which did not come immediately within his own experience. But, for the most part, the information is derived from direct observation during a period of about thirty years.

The authorities referred to are, it is hoped, fairly quoted throughout; at least it has been the desire of the Author not to assert as of his own observation that which is due to the statements of eminent engineers.

LEOMINSTER.
May, 1888.

CONTENTS.

SECTION

I. *EMBANKMENTS OF WATERWORKS RE-SERVOIRS.*—Consolidation in various degrees by different means of raising the bank 1

II. *QUANTITY OF WATER TO BE STORED.*—Part only of the rainfall can be dealt with; the dry-weather flow; height of bank; depths of reservoirs 16

III. *APPURTENANCES OF RESERVOIRS.*—Bye-channel; waste-weir; gauge-basin; valve tower 26

IV. *DISCHARGE OF WATER FROM THE RESERVOIR.*—Pipes; culverts 31

V. *APPROXIMATE COST OF A STORAGE RESERVOIR.*—Work to be done. 35

VI. *CONCRETE FOR EMBANKMENTS AND DAMS.*—To be solid or not according to its intended use; washing the materials; interstices of the materials 45

VII. *STREAM GAUGES.*—Fixing the gauge; construction of a temporary gauge; co-efficients of discharge derived from Mr. Blackwell's and Mr. Simpson's experiments; the co-efficient of Mr. J. B. Francis; depth of water measured from a still head compared with the depth upon the outer edge of a weir 51

CONTENTS.

SECTION	PAGE

VIII. *RAINFALL.*—Increases from the West to the gaps in the main range of hills, then decreases Eastwards; Mr. Symons's records; rain-gauges   64

IX. *AREAS OF RIVER-BASINS.* — South coast; East coast; North-east coast; North-west coast; West coast   72

X. *CONDUITS.*—Degree of fulness; form of channel; open and covered conduits; pipes over ravines; air in pipes   78

XI. *TUNNELS.*—Cost of several waterworks tunnels   90

XII. *SERVICE RESERVOIRS.*—Their position and general construction   92

XIII. *PRESSURE AND ITS EFFECT IN PIPES.*—Head of water; proof, from observation, that the head is proportional to the square of the velocity; hydraulic gradient; strength of cast-iron pipes   97

XIV. *AQUEDUCTS.*—Large conduits proposed for the supply of London some years ago   113

XV. *RIVERS AND WATERCOURSES.* — Various formulæ of the mean velocity; the bottom velocity deduced from the observed surface-velocity   120

XVI. *COMPENSATION TO MILLS.* — Weir for separating clear water from flood waters; example of a millowner's reservoir; value of water-power   127

XVII. *OF WATER-POWER IN GENERAL.*—Basis of calculation of horse-power   135

CONTENTS.

SECTION		PAGE
XVIII.	*WATER WHEELS.*—Various kinds of wheel— Overshot, high breast, low breast, undershot, current wheels; flow of water from openings; speed of wheels; force of water-current	139
XIX.	*CORN MILLS.*—Relative velocity of water and of wheel; quantity of water used for grinding corn with various forms of wheel	168
XX.	*WORK DONE BY WATER WHEELS.*—Pumping water; grinding saws, &c.; working forge hammers	184
XXI.	*TURBINES.*—Elementary form; high and low falls; action of water on various forms of turbine	192
XXII.	*DOMESTIC WATER-SUPPLY.*—Quantity used for various purposes	217
XXIII.	*SERVICE RESERVOIRS.*—Situation; puddle lining; banks of sand	222
XXIV.	*DISTRIBUTION OF WATER.*—Variable flow during the day; watering roads; meters; preventing waste	230
XXV.	*PUMPING-MAINS AND ENGINES.*—Standpipes; Cornish engines; rotative beam engines	241
XXVI.	*FLOODS.*—Lowering the water-level of rivers; weirs on rivers; flow off the ground; Thames and Severn compared; effect of repeated floods	247
XXVII.	*STORAGE OF FLOOD WATERS.*—Not practicable in the lower reaches of a river; regulating dam and storage reservoir separate and combined	263
XXVIII.	*RELIEF OF LAND FROM FLOODS.*—Cost per acre; rating lands	271

SECTION	PAGE
XXIX. *REGULATION OF FLOOD WATERS.* — Heavy falls of rain; quantity flowing off the ground; reservoir capacity of various sites	278
XXX. *RIVER CONSERVANCY.* — Difficulties of the subject	287
XXXI. *COUNTY BOARDS AND WATERSHED AREAS.*—Various remarks pertaining to the subject	295

WATER ENGINEERING.

SECTION I.

EMBANKMENTS OF WATERWORKS-RESERVOIRS.

RESERVOIR embankments in general are made with the materials found on or near the site, be they earth, shale, or stone. With earth or shale the English practice makes the watertightness of a reservoir dependent on a wall of puddled clay, carried up in the centre of the embankment from retentive ground below its seat. The thickness of the puddle wall varies from 4ft. to 8ft. at the top, and at the bottom it is as much more as is required for a batter on each side of an inch to every foot in height. But it is not good practice to make the thickness of the puddle wall at the top so little as 4ft., and 6ft. is the least dimension that ought to be given to it, increasing beyond this, according to the quality of the clay, to 8ft. in perhaps most cases. Thus, where the height of the bank is 24ft., the thickness of puddle at the bottom would be 12ft., or half the height. Where the height is 48ft., the thickness would be 16ft., or one-third of the height; and it has been recommended to make the thickness one-third of the height also in high banks, and this would be, for a bank 96ft. in height, 32ft., although, according to the proportions above stated, it would be 24ft., or, in this case, one-fourth of the height.

Puddled clay, upon which the watertightness of reservoir embankments has for so long a time depended, since, at least, the time of Brindley and canal-making, has come to be regarded with disfavour and distrust where it is itself exposed to the action of water or of the air. When some canal-makers complained to Brindley that they could not stop a run of water, his advice was to "puddle it, puddle it"; and the good reputation which puddle has always had in waterworks-engineering led at one time to its careless use, and it was solely depended upon for the watertightness of reservoir embankments, which, however, was not always effected.

Certainly, some reservoir embankments have been made of very great height, and they have stood safely, and there is no apprehension whatever that they will not always continue so, with, of course, due attention to the outer parts, to repair the continued action of the weather—frost, thaw, rain. But these embankments have been well-made, not only in the core, which is formed of puddled clay from end to end of the bank, like a wall, but on each side of it; on the inside to prevent the water in the reservoir having access to it in any considerable body, and on the outside to prevent the water with which the puddle is made being drawn out of it, whether by evaporation or capillary attraction, by which it might become gradually dried, and the cracks which would be consequently formed might extend so far into the wall as to reduce its virtual thickness materially. Rough stone, for instance, or earth loosely tipped in, would admit the air to it, and the water incorporated with the puddle would continue to evaporate as long as the air itself was not saturated with moisture.

Puddled clay does not readily part with the water with which it is incorporated, and not at all if the outer and drier air is wholly excluded from it; but this can only be done by placing against it a compact material. On the other hand, in the presence of water having any motion, it easily dissolves—melts away; and the best puddle for a wall—that is, the closest in texture—melts the soonest under this action; nevertheless, it is absolutely necessary that puddle should be so worked as to bring it to a close texture; but at its best it is porous, and if water has access to it under great pressure, it is only a question of degree how much water will be forced through the puddle wall. In the first place, a great deal of water is used in making it; dry clay will absorb nearly a third of its own weight of water; and if it be taken, not in a purposely dried state, but in a naturally dry state, it will absorb, in the process of puddling, say an eighth or a sixth of its weight of water. Thus, water acting under pressure on one face of a puddle wall tends to force water out on the other side, being itself incompressible, and the quantity it can force through in a given time is limited only by the smallness of the pores and the length and tortuousness of the course which a run of water must follow. It is, therefore, highly important to protect the puddle from contact with water under great pressure. It might almost seem that if it requires so much protection it can be of little use, and that it might be dispensed with altogether; but this extreme would be as unwise as that of depending solely upon it without precautions; and as it is so readily procurable, it remains still a necessary material for waterworks-reservoirs.

There are two kinds of puddle—clay puddle and gravel puddle—proper to be used for different purposes, or rather in different positions for the same purpose of watertightness; the one consisting of clay only, the other with stones incorporated. When there is a liability to a wash of water, stones are necessary to hold the clay together, and of stones rounded gravel-stones are the best, besides being generally the more easily obtained, inasmuch as they are more uniformly dispersed through the mass of clay than angular stones can be in the process of working, which is by cutting and cross-cutting with long-bladed tools, reaching down at every stroke, through the layer of clay being worked, into the puddle beneath it. The tool slips past a rounded gravel-stone without much disturbance of the adjoining clay; but angular stones, when struck by the puddling tool, have a tendency to congregate together.

Gravel puddle is proper to be used where weights have to be sustained, as for the walls of a service reservoir. Clay puddle can hardly be worked stiff enough to support the weight, which squeezes the clay outwards and upwards, and there is an inconvenient settlement of the wall; but with gravel puddle the weight can be sustained and water-tightness effected also, because with a mixture of gravel the puddle can be worked stiffer.

The sketches given below which are numbered 1, 2, and 3, show respectively a longitudinal section, a plan, and a cross-section of an embankment formed of earth, with a puddle-trench and wall. The trench is shown to be cut straight down, so that the bottom is as wide as the top, and the advantage of this over sloping sides is that, in case the ground should not prove watertight at the depth anticipated, the trench may be carried down to any further depth. The trench is filled in with puddled clay up to the surface of the ground, above which, up to and above the top-water level of the reservoir, the puddle is continued as a wall, having a batter on each side of 1in. to a foot, and finishing 8ft. thick at the top; 4ft. is not a sufficient thickness, 6ft. is sufficient with precautions, but 8ft. is advisable; if, however, there be the means of placing a nearly watertight mass of earth next the puddle, as B B on the sketch, the thickness of the top of the puddle may be 6ft.

In the longitudinal section, Fig. 1, the bottom of the

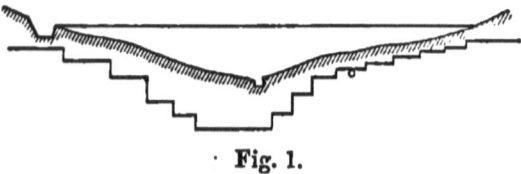

Fig. 1.

puddle-trench is cut into the hill-side as far down as to pass through loose or fissured ground to a water-tight bottom, which is often found nearer the surface on one side of the valley than on the other. That side should be chosen for the discharge-pipe or culvert, but it cannot always be determined before the puddle-trench has been cut.

On the plan, Fig 2, A B is the original course of the

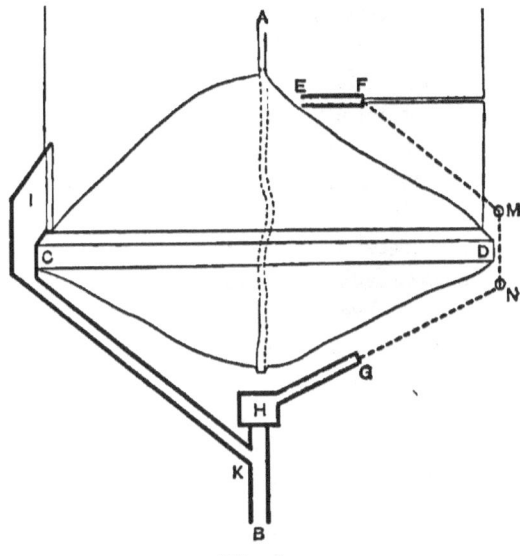

Fig. 2.

stream; C D, the two ends of the embankment; E F, an open channel, cut near the bottom of the reservoir; F G, the discharge-pipe, laid in the solid ground; sometimes laid in a culvert of masonry, itself laid in the solid ground. G H, an open channel; H, the gauge-basin; I, the waste-weir; I K, the bye-wash.

The discharge-pipe has a valve on its inner end, at F, and it is necessary to erect a tower of some kind, from the top of which it can be worked. Access to the top of the tower is procured by means of a bridge, either from the top of the embankment or from the side of the reservoir, as shown in the sketch. When the puddle-trench is everywhere sunk below the level at which the water is drawn off, the discharge-pipe is taken round the end of the bank, instead of direct from F to G, for the sake of being in the solid ground, and a tunnel is then preferable to an open trench, as shown by the dotted line F, M, N, G, shafts being sunk at M and N. It may not be quite justifiable to put the preference of a tunnel on the less disturbance of the ground, for with careful timbering, perhaps less disturbance would occur with an open trench. The open trench can, at least, be closely filled in again, which cannot be said with certainty of the space between the crown of a tunnel and the earth above it.

Although tunnels have of late been much advocated instead of culverts, it would certainly be unwise to adopt them in every case. In every case, however, there is a necessity for a tower or valve-well, for a valve cannot be satisfactorily worked with sloping rods, and because it is absolutely necessary to place a valve on the inner end of a discharge-pipe, if any regard at all be had to the safety of the reservoir. Certainly, discharge-pipes have been laid under embankments, which have had no valves at the inner ends, the discharge being controlled at the outer ends, at the foot of the embankment; but considering that a cast-iron pipe, if of large size, may be broken by the external pressure of the earth, and that, whatever the size, it is subject to leakage at the joints, where it cannot be got at for repair, this method will not bear repetition. A valve, then, being necessary at the inner end of the discharge-pipe, and a tower for access to it, it becomes desirable to draw off the water, not from the bottom of the reservoir, where it is muddy, but as near the top, at all times, as is practicable, and the tower affords facilities for the insertion of a short pipe through the wall at as many different heights as may be desired.

Instead of placing the tower or valve-well in the reservoir at or near the foot of the slope of the embankment, it has been sometimes placed almost in the middle of the bank, close to the puddle wall, and the water is conducted to it by a culvert from the foot of the inner slope of the embankment; but this position does not afford the facilities of drawing off the water at different heights that the exposed tower does.

With a good masonry culvert or tunnel of sound bricks or thick stonework, a pipe in addition seems superfluous, where the reservoir is constructed for one discharge only; but if it be so situated that two discharges are required from it, one, for instance, for a town supply, and another for the stream itself, the continuation of the supply-pipe up to the valve-tower, through the culvert or tunnel, is necessary, the discharge to the stream running on the bottom of the culvert or tunnel beneath the supply-pipe, which is supported from side to side so as to leave room beneath it for the stream discharge. The quantity given to the stream, under these circumstances, is usually required to be open to inspection by turning it over a gauge-weir to a depth agreed upon, the length of the gauge being fixed, by which the quantity being discharged can be ascertained for the satisfaction of millers and others interested in the stream.

The waste-weir and bye-wash are shown in the sketch on the opposite side to the other works, but, according to circumstances, they may be on either side. They are very important parts of the work, for when the reservoir is full at a time of flood nearly all the water must pass that way, and therefore not only must the length of the waste-weir be sufficient to prevent the flood rising to a dangerous height against the embankment, but the bye-wash must be so constructed that it will not be torn up by a great body of water rushing down so steep a channel.

In the cross section, Fig. 3, A is the puddle-trench and

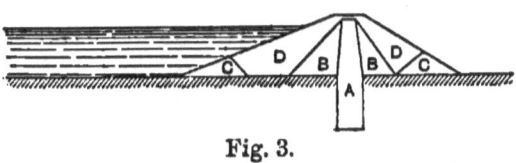

Fig. 3.

wall; B B, material selected from that excavated as being the most easily compacted; C C, a rough stone bank

8 EMBANKMENTS OF WATERWORKS-RESERVOIRS.

necessary for the outer slope, and advisable also for the inner; D D, the rest of the material excavated, except peat. This is a material which, although sometimes abundant on reservoir sites, is very objectionable in an embankment, even in small quantities.

The top of the embankment, when consolidated, has usually been made from 4ft. to 6ft. above the top water level, and in situations much exposed to high winds not less than 6ft.; for a gale of wind blowing down the reservoir rolls against the face of the bank waves of considerable height, and this may occur when the reservoir is more than full, and water going over the waste-weir at a depth of, perhaps, 2ft., thus reducing the clear height to 4ft.

The width of the top of the bank has usually been made from 12ft. to 24ft. The proper width is determined by the considerations of whether or not a roadway is required to be made along it, and what width the puddle wall is to be at the top. Leaving out of consideration the question of roadway, which must be always a local requirement, the width necessary to properly enclose and protect the puddle wall is about twice the width of the top of the puddle wall itself.

The safety of a reservoir embankment does not depend upon one thing only, as upon the perfection of the puddling, but equally upon several others, chief amongst which is the consolidation of that part of the bank which lies within the puddle wall, and into which, therefore, water would penetrate if not prevented, and this prevention depends chiefly upon the thickness of the layers of earth with which the embankment is raised. Thin layers are always desirable; but the maximum thickness which may be allowed depends on the material and the means by which it is deposited in the bank. A heavy and dry material may be deposited in thicker layers than one of less specific weight, with equal effect in both cases in forming a solid bank. The thickness allowable has been variously stated at 6in., 12in., 18in., and 2ft.

Where the chief consideration has been the urgent progress of the work to meet a demand for water, as was the case when, in order to meet the demands of the approaching summer the embankment of the reservoir on the Dale Dike at Bradfield was urged forward by the Sheffield Waterworks Company, in the early part of the year 1864, and which burst in March of that year; the earth forming the bank on each side of the puddle wall had been tipped from rail-waggons at a height of 5ft. or 6ft., and more; but few waterworks engineers would sanction that, except for the outer part of the bank, and even for that part it is too high, because the large and small rock are not evenly mixed; the large lumps roll to the bottom of the tip and form a layer there of themselves; the finer stuff lying together upon the coarse layers; whereas, to prevent slipping, which is one great object of all the precautions, the large and the small should be as evenly mixed as is practicable, except in those parts where material is purposely selected to be deposited, fine in one part and large in another, for different purposes. If the material tipped into the bank were all stone, the rolling of the large lumps into a separate layer would be less objectionable; and, indeed, with stone for the material it might even be an advantage that the layers should be so arranged, for the one would drain the other, and the bank, as a whole, would retain less water than it otherwise would do; but with other materials, such as shale, whether blue or dark, or, indeed, any material but stone, the same object would not be effected.

In making the inner part of the embankment of a reservoir the object is to make it as compact as possible to prevent the water reaching the puddle wall, and, therefore, the material should be small and clayey; but in the outer part of the bank dryness and stability are the chief objects, for which larger and harder material is more suitable. The inner slope of an embankment is usually made flatter than the outer slope, as 3 to 1 inside and 2 to 1 outside, for the inner part of the bank is more liable to slip than the outer. When the water-level in the reservoir is reduced, if the inner part of the bank has been penetrated by water and become partially saturated, slips are more likely to take place, and the flatter slope meets this tendency; but outside, with dry materials, 2 to 1 is as good a slope as 3 to 1 inside. As a further precaution against slipping it is desirable to keep up the outer parts, next the slopes, higher than the inner parts next the puddle wall, as in Fig. 4.

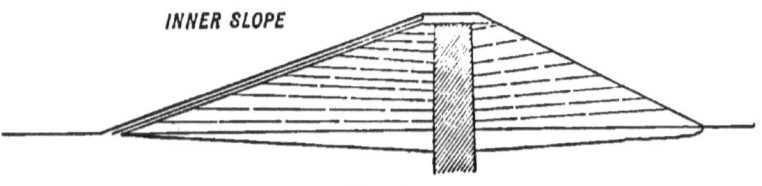

Fig. 4.

If rail-waggons be used at all, it is desirable that they should be of small size, to hold, say, not more than one cubic yard; but if waggons be allowed, there will always be a tendency on the part of those who do the work to make the tips high and the layers thick, to avoid much shifting of the rails if possible, and especially by the use of side-tip waggons; but if the instruments for carrying the stuff into the bank be confined to barrows, dobbin-carts, and common one-horse carts, it does away with this tendency in a great measure. When large waggons are advocated, it is asserted that the greater weight and the heavier rails compensate for the greater height of the tips, and so an equally good consolidation of the earth is effected. The degree of consolidation, by the weight of the instrument by which the earth is transported, admits of proof. In respect of any one square yard near the middle of the bank, lengthwise, the consolidation of the one immediately preceding it may be shown as follows:—Premising that whatever the thickness of the layers, variation in weight of the different kinds of earth, or time of construction, the degree of consolidation by the weight of the carrying instrument must be in the inverse ratio of the weight carried each journey, the weight of the instrument being reckoned twice, going full and returning empty.

Thus, if w = weight of earth carried, and W = weight of instrument carrying it,

$$\frac{W + w}{w} = \text{degree of consolidation.}$$

The actual weights of different kinds of earth vary much; but, for the purposes of comparison, an average weight of 2600lb. per cubic yard may be taken, or 96lb. per cubic foot, and the effect of the carrying instrument may be stated thus:—

BARROW	65lb.
Proportion of length of plank road	15
Wheeler	150
Road-shifter	140
Earth, 1⅔ c. ft. at 96lb. per c. ft.	160
	530
Returning empty	370
Mean consolidating weight	450lb.

$$\text{Degree of consolidation} = \frac{450}{160} = 2\cdot 81.$$

The quantity of earth which a dobbin-cart will hold may vary from ¾ cubic yard to 1¼ cubic yard, and may be taken at an average of 1 cubic yard, and its weight at 10cwt.

DOBBIN-CART	1,120lb.
Horse, 10 cwt.	1,120
Driver	130
Two Tipmen	280
Earth, 1 c. yd.	2,600
	5,250
Returning empty	2,650
Mean consolidating weight	3,950lb.

$$\text{Degree of consolidation} = \frac{3,950}{2,600} = 1\cdot 52.$$

```
COMMON CART, 12 cwt. .. .. .. .. ..   1,344lb.
Horse, as before   .. .. .. .. .. ..   1,120
Driver .. .. .. .. .. .. .. ..           130
Two Tipmen     .. .. .. .. .. ..         280
Earth, 1 c. yd. .. .. .. .. .. ..      2,600
                                       ─────
                                       5,474
Returning empty.. .. .. .. .. ..       2,874
                                       ─────
Mean consolidating weight   .. .. ..   4,174lb.
```

$$\text{Degree of consolidation} = \frac{4,174}{2,600} = 1\cdot 60.$$

Rail-waggons holding one cubic yard are very handy things for shifting earth. If the gauge of the rails be 3ft., and the waggon held one cubic yard, its weight will be about 1000lb. The weight of a proportionate length of rails and sleepers will be about 120lb., and the statement will stand thus :—

```
SMALL RAIL-WAGGON .. .. .. .. ..       1,000lb.
Proportion of length of rails and sleepers  110
Horse .. .. .. .. .. .. .. ..          1,120
Driver .. .. .. .. .. .. .. ..           130
Two Tipmen   .. .. .. .. .. ..           280
Earth, 1 c. yd. .. .. .. .. .. ..      2,600
                                       ─────
                                       5,240
Returning empty.. .. .. .. .. ..       2,640
                                       ─────
Mean consolidating weight   .. .. ..   3,940lb.
```

$$\text{Degree of consolidation} = \frac{3,940}{2,600} = 1\cdot 51.$$

The consolidating effect of a two-yard waggon is less, thus :—

```
TWO-YARD WAGGON .. .. .. .. .. ..      1,600lb.
Proportion of length of rails and sleepers  140
Horse, 11 cwt. .. .. .. .. .. ..       1,232
Driver .. .. .. .. .. .. .. ..           130
Three Tipmen  .. .. .. .. .. ..          420
Earth, 2 c. yd. .. .. .. .. .. ..      5,200
                                       ─────
                                       8,722
Returning empty   .. .. .. .. ..       3,522
                                       ─────
Mean consolidating weight   .. .. ..   6,122lb.
```

$$\text{Degree of consolidation} = \frac{6,122}{5,200} = 1\cdot 18.$$

INSTANCES OF FAILURE.

These differences of degree are small when compared singly; but when it is considered how often they are repeated, the difference becomes very great in the whole. All other reasons for making the inner part of an embankment very compact are strengthened by the consideration that when it is so, the pressure of the water in the reservoir takes effect upon the bank in a direction perpendicular to the slope; but if water penetrates it the pressure is horizontal.

Within the century two reservoirs have burst—viz., Bradfield and Holmfirth—and Mr. Bateman* referred to these in a discussion on Tunnel Outlets at the Institution of Civil Engineers, and said that the Holmfirth reservoir which burst in 1852 was never filled except on the single occasion on which it burst. The water escaped through the fissures of the rock on which the embankment was constructed, and gradually washed it down in such a way that the top of the embankment was lower than the wasteweir, and when an extraordinary flood came it passed over this low part of the bank and carried the whole away. The Bradfield embankment was constructed on ground liable to slide down upon the face of an underlying flag rock with a smooth surface. But, besides this, the inner part of the bank was of very loose material, and the pressure would be horizontal, the puddle wall and the outer part of the bank having to bear the whole pressure of the water in the reservoir because the inside slope was permeable by water. The puddle trench was sunk through the flag rock into the shale below, but there, on the outside, lay the flag rock at an inclination of 1 in 4 as smooth as polished marble, and when the pressure came against the puddle wall the whole thing gave way.

That was the opinion of Mr. Bateman given in 1879. He was on the spot immediately after the accident, but did not then give any opinion at the local inquiry. Mr. Rawlinson's opinion was different. There were two lines of 18in. cast-iron discharge-pipes laid side by side in puddle in a trench excavated in the rock, but above

* J. F. Bateman, Esq., F.R.S., M.Inst.C.E.

the bottom of the main puddle trench where the pipes crossed it, and Mr. Rawlinson's* opinion was that the joints of the pipes were opened by reason of the surrounding puddle sinking in the middle, and allowing water to creep along outside the pipes and so make a beginning of a more serious run of water, as there were no valves on the inner ends of the pipes, and if such a beginning was made the run of water could not have been controlled, the valves being at the foot of the outer slope of the embankment. Mr. Beardmore, who together with Mr. Rawlinson conducted the inquiry, concurred in this opinion.

But there is another way in which the water may have been let out of the reservoir, and for the purpose of stating it the construction-in-chief of the embankment may here be recapitulated.

It was 95ft. high, had a top width of 12ft., and slopes $2\frac{1}{2}$ to 1 both inside and out. The puddle trench was sunk to the depth of 60ft. to watertight ground. The thickness of the puddle wall was 18ft., diminishing to 4ft. at the top, but it was very well made. The earth, however, forming the bank on each side of it was tipped from rail-waggons at a height of 6ft., and more. The breach in the bank was 100 yards wide and 70ft. deep, and 90,000 cubic yards of earth were carried away by the water in thirty or forty minutes out of a total quantity in the embankment of 406,000 cubic yards. This breach was near the middle of the length of the embankment. The portion left standing, on the north side, presented this appearance a few days after the accident :—

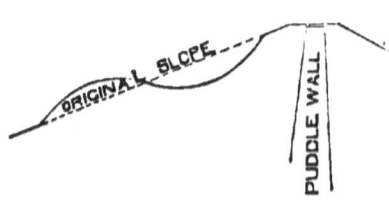

Fig. 5.

Now, considering that the place where this appearance was presented was fifty or more yards from the middle of the breach, the slip which was here apparent — the tail end of it as it were—was, probably, a very great one at or about the middle of the

* Sir Robert Rawlinson, C.B., M.Inst.C.E.

bank, and sufficient to withdraw from the inside of the puddle wall the whole of the earth which supported it against the pressure of the outer part of the bank. If this earth slipped away from the puddle wall for any considerable length and depth, it would almost certainly follow that the top of the puddle wall would fall inwards, and if only a few feet in height of the upper part of the puddle wall fell in it would let the water begin to flow out and down the outer slope of the bank, and, of course, the action would go on with quickly-increasing effect, until, in a short time, it would seem to go all at once.

But whatever inference may be drawn from it, the slip was a fact to the extent shown in the sketch above. Now what was the cause of the slip? The material of the bank was chiefly shale—the dark shale of the coal measures—but this is rather a good than a bad material for the purpose, being largely composed of thin, indurated beds of rock. The slip then probably occurred by reason of the large quantity of water which found its way into the bank, the upper portion of it having been raised quickly for the purpose before stated. In the lower portion of an embankment the area is so large that the earth is almost necessarily spread out in thin layers, and the bank is therefore raised by slow degrees, but near the top, as the width diminishes, progress upwards is more rapid.

SECTION II.

Quantity of Water to be Stored.

The capacity of a reservoir to control the flow of water from a given area of ground used formerly to be calculated on the basis of a proportion of the average rainfall, as two-thirds, one-third being deducted for loss by evaporation and absorption, but it was found by further experience that the loss could not in all cases be taken as any certain proportion of the average rainfall, although in many cases one-third was a near approximation. In the case of reservoirs the capacity of which had been calculated on the basis above named, it was occasionally found that some of the water came down to the reservoir when it was full or nearly so, and passed over the waste-weir beyond control, and, consequently, that instead of the reservoir being full at the commencement of a drought, it was sometimes considerably below the top-water level at such times, and that the quantity stored did not keep up the required daily supply for a sufficient length of time.

The storage capacity of a reservoir large enough to equalise the flow of water over a long drought, or two or three shorter ones with intervening winter rains of less than the usual amount, and so that the annual rainfall is made to yield its daily average quantity throughout long periods of time, can be found when the past rainfall daily for a long time has been ascertained, the number of dry days in the future being assumed to bear the same relation to wet ones as they have done hitherto, and that the past average yearly quantity of rain will continue the same;

then the capacity of any reservoir may be found, being measured by the longest period of defect of supply into and discharge out of the reservoir.

The rainfall of a long series of years must be considered in connection with that of a short series of dry years, as three or four. When a table of the rainfall of any locality, extending over a long series of years, is examined, it is found that the mean depth of three consecutive dry years is about one-sixth less than the general average. The wettest year has a rainfall about half as much again as the general average, the driest year one-third less, and the mean of the driest three consecutive years one-fifth less, than the general average.

Thus in some years a great deal of water passes off the ground by the streams to the sea which cannot be stored. Mr. Hawksley * stated before the Royal Commission on Water Supply, which sat in the year 1868, that gaugings of the actual quantity of water going over the waste-weirs of reservoirs show that it is about one-sixth of the quantity due to the average rainfall, and this agrees with deductions from the tables of rainfall, showing that reservoirs of the capacities of those which permitted one-sixth of the general average rainfall to pass over the waste-weirs do not deal with more than the average depth of three consecutive years of least rainfall. In these cases therefore one-sixth of the average must first be deducted, then the loss by evaporation, &c., and the depth remaining will represent the available quantity upon which the capacity of a reservoir may be calculated, according to the number of days' storage required.

A great portion of the water yielded by light rains is absorbed into the ground, and neither flows directly off the ground by streams nor sinks far into it to issue again in springs, but is evaporated, partly from the surface of the ground itself, and partly from the leaves of the vegetation into which it enters, only a small part being left in the vegetation itself, the greater portion of the

* T. Hawksley, Esq., F.R.S., M.Inst.C.E.

water which enters into it being evaporated from its multitudinous surfaces.

The annual depth of rain which in this way does not contribute to the streams varies with the nature of the ground on which it falls, being least on a non-absorbent and precipitous area such as mountains of slaty rocks of old geological formation. The longer the time the water takes to reach the reservoir the more of it will be lost, and besides the difference caused by the declivity of the ground the quantity evaporated depends on the humidity or dryness of the air of the locality, and this latter condition varies much in different parts of the country. The depth of rainfall which does not contribute water to the streams at the sites of reservoirs varies from 9in. to 18in. in different parts of the country.

The number of days' storage a reservoir for the water-supply of a town should contain varies with the average annual depth of rain and also with the frequency with which it falls. Mr. Bateman, in his evidence before the Commission already named, said that whereas 120 days' storage would be sufficient on the western side of the country and some other parts, where the rain falls on a great number of days in the year, 240 days' storage would be required on the eastern side of the country, where the rainfall is both less in depth and less frequent, and it has been estimated that in certain situations on these sides of the country respectively 150 and 300 days' supply would not be too much.

For the sake of illustration we may take an instance where the average annual rainfall is 42in. Deducting one-sixth, 35in. would be the depth to be reckoned upon, and supposing the depth allowed for loss by evaporation, &c., to be 14in. the available depth would be 21in.; and, in the absence of exact measurements in every part of the area from which the water from time to time flows, that depth of 21in. must be supposed to extend over the whole area, if it is not of inordinate extent.

Where the area is very extensive, that depth of rainfall may not extend over the whole of it, and this may have been one of the reasons why expected quantities of water have sometimes not been yielded by a drainage area of given extent; but it is the nearest approximation to the truth which can be made, and some concession towards it may be considered to be made in deducting one-sixth from the known average rainfall of a few places within the area, although in some cases that quantity has been found to pass actually over the waste-weirs, thus furnishing a direct proof that with such reservoir-room as has been there adopted that part of the water could not be stored.

When, as is the case sometimes, large areas are dealt with, the available quantity of water is conveniently reckoned per thousand acres of the watershed area. Upon 1,000 acres an available depth of 21in. would yield an average daily quantity of

$$\frac{1,000 \times 43,560 \times 1 \cdot 75}{365} = 208,800 \text{ cubic feet, of which}$$

about two-thirds might be used for supply to houses, manufactures, or for power away from the stream, or for irrigation, leaving one-third or thereabouts to the stream itself; or rather, this latter would be first apportioned, leaving two-thirds for other purposes, and the apportionment would stand thus—

One-third = 69,600 cubic ft. per day to the stream.
Two-thirds = 139,200 „ „
$6\frac{1}{4}$
___

870,000 gallons per day for supply, and so on per 1000 acres, within limits defined by our knowledge of the real average rainfall over any area, the certainty of the calculated average increasing with the number of places of observation over a given area, but becoming too uncertain for use when the area is very large and the places of observation few.

The quantity above set down for the stream would

render it more powerful in dry weather than it would be without a reservoir. The average flow of the stream in this case during dry weather would probably not be more than two-thirds of the quantity above stated, for measurements of the dry-weather flow of streams show about 30 cubic feet per minute per 1000 acres of the drainage area.

This, no doubt, varies with the nature of the ground from which the water flows, being greater or less as the ground is more or less absorbent. Where the ground in the upper and middle portion of a river-basin consists of chalk, oolite, or sandstone the dry-weather flow forms a larger part of the whole quantity flowing off the ground than where it is of hard and impervious rocks, with the steeper surfaces which always accompany those harder rocks; and the kind of ground which affords a medium dry-weather flow between these extremes is that which consists of alternate formations of sandstone and shale, or of oolite and lias clays, in either case covered in parts with clay and gravel.

The rainfall observations, if sufficiently numerous, will show on an average of years that where the annual quantity is great the number of rainy days is greater than in those parts of the country where the rainfall is comparatively small [this is not, as it might seem to be, a necessary consequence], and so, although more water per day is discharged from a reservoir per 1000 acres of its drainage area in the former situations than in the latter ones, the loss of water is oftener replenished, and in proportion to the frequency of this action the capacity of the reservoir may be smaller.

The experience in England during the last twenty-five or thirty years has been that, in order to equalise the flow of water day by day during three or four consecutive years of least rainfall, the capacity of a reservoir should be sufficient to yield about 120 days' supply where the rainfall is frequent, and 240 days' where it is infrequent; but it is here assumed that the reservoir is full at the commencement of a drought, whereas it may not be so if the previous winter's rain has not yielded a surplus over the regular supply sufficient to fill up the reservoir completely, and in this view the capacity should be increased perhaps to 150 days' supply in the one case and 200 days' in the other.

From an examination of Mr. Symons's [*] annual records of rainfall it appears often that nearly as many rainy days occur in the southern and eastern counties as in the western and midland counties, although the rainfall in these is much greater; but the days set down as rainy in the records include all those on which one-hundredth of an inch of rain has fallen, and this is so little that it often adds nothing to the run of the streams. The rainy days, the frequency of which affects the capacity of storage reservoirs, are not numerous.

If in the example where 21in. is the available annual depth of water over the initial area of 1,000 acres, the proper number of days' supply were, from its position, say 180, the capacity of the reservoir would be

$$208,800 \times 180 = 37,584,000 \text{ cubic feet.}$$

Deducting the dry-weather flow due to 1,000 acres at the rate of 30 cubic feet per minute, that is 7,776,000, there remains 29,808,000, or say 30 million cubic feet, as the capacity of the reservoir, which is at the rate of 30,000 cubic feet per acre of the watershed area.

Whether the dry-weather flow of the drainage area should be deducted from the capacity of a reservoir in the manner already stated, will depend upon the method by which the number of days' storage is arrived at; whether upon the basis of (1) the rainfall table, showing the absolute length of time during which no rain has fallen; or (2) upon the stream gaugings, showing the daily quantity reaching the site of the reservoir; or (3) upon the difference between two quantities, one of which is the quantity of water in an existing reservoir at the commencement of a drought, to which is to be added the quantity coming into the reservoir during the drought,

[*] G. J. Symons, Esq., F.R.S., F.R.Met.Soc.

whether from springs or casual rainfall—and the other quantity is that which has been discharged during the drought, as ascertained by the daily sinkings of the water-level of the reservoir, in which case it is evident that the total quantity is dealt with on either hand, and, therefore, no deduction can properly be made.

But this method is true only of that reservoir, unless the dry-weather flow, where it is proposed to make a new one, is the same, area for area; whereas by the other method the rule becomes general, and is applicable alike to cases where the dry-weather flow is large, comparatively, and where it is small. It is, therefore, better as a method, although the other is more exact for a particular case already existing. The dry-weather flow of the stream at the site of the reservoir is made up of the numerous small springs issuing within the drainage area, produced because of the absorbency of the ground during wet weather, and its yielding of the absorbed water gradually, and to this extent of the dry-weather production the drainage ground is essentially a part of the reservoir, and for this reason the quantity should be deducted from what would otherwise be the necessary capacity.

Were the ground wholly impermeable there would be no run of water in dry weather, and, in that case, the capacity of the reservoir itself would need to be so much larger; but where, as is assumed in the example, there is a dry-weather flow which, on the average, during a drought, is equal to 30 cubic feet per minute per 1,000 acres, that quantity may be deducted from the calculated storage-room in order to arrive at the actual capacity of the reservoir. The quantity so deducted belongs to the reservoir, and is, in fact, part of the yield reckoned upon in the available depth assumed, as, in the case adduced, 21in., and whether it be conducted past the reservoir and form part of the supply, or it run through the reservoir and form part of the discharge into the stream, makes no difference in the reservoir-room proper to be provided;

HEIGHT OF BANK.

but where the number of days' supply to be stored is founded on the length of a drought as ascertained from the rainfall returns, the dry-weather flow ought to be deducted. But the formation of the ground on reservoir sites varies much, so that any rule of this kind gives only an approximation to results in particular cases.

The height of the top of the embankment above top-water level is often 4ft. in small reservoirs and 6ft. in large ones. Where the depth of water at the site of the embankment does not exceed about 30ft., the top of the bank may usually be 4ft. above top-water, or the level of the waste-weir. With a depth of from 30ft. to 50ft., the top of the bank may be 5ft. above that level, and where the depth is 60ft. and upwards, the height should be 6ft.; but it somewhat depends upon the direction in which the reservoir lies lengthwise, and to its exposure to gales of wind. Irrespective of this, a certain height is necessary to meet a flood which may raise the water-level above the waste-weir, and as a matter of prudence and safety the waste-weir should be of such length as to prevent the water rising, on the occurrence of the greatest flood, to a height more than about 2ft. above the waste-weir. But beyond that, an allowance must be made for the further height to which the water may be driven by wind, and the greater the depth of water at the embankment the longer will the reservoir be on the same site, and on a long reservoir the water is driven up by the wind to a greater height than on a short one.

In the upper parts of river-basins where reservoirs may be made, the opposite hill-sides approach each other towards the bottom, and almost meet in the stream with straight slopes. This is a general characteristic; but a really good site for the formation of a storage reservoir widens out in the bottom above the site of the embankment, and the only way of ascertaining the quantity of water the site is capable of affording, with any given height of bank, is by actual measurement, either by sections across the valley or by contour lines laid down

24 QUANTITY OF WATER TO BE STORED.

upon a plan every 4ft. or 5ft. in height; but we may estimate approximately what quantity of water a fairly good reservoir site would probably afford by examining sites the capacities of which have been ascertained, and comparing these with the respective heights of the embankments, or, rather, with the greatest depths of water, D, and their areas, A, taking these also at several different heights on the same site. The results of an examination of 75 such cases where the depth D varies from 20ft. to 80ft., both inclusive, are as follows:—In each case the cubic capacity is divided by the area of the water surface, giving the average depth; this then is divided by the greatest depth, D, giving the ratio set down opposite each case.

No.	Ratio.	No.	Ratio.	No.	Ratio.	No.	Ratio.
1	·612	20	·49	39	·452	58	·4
2	·583	21	·487	40	·45	59	·4
3	·580	22	·485	41	·45	60	·4
4	·572	23	·473	42	·447	61	·4
5	·560	24	·468	43	·435	62	·4
6	·543	25	·468	44	·431	63	·396
7	·537	26	·468	45	·427	64	·395
8	·528	27	·466	46	·422	65	·394
9	·526	28	·466	47	·42	66	·38
10	·525	29	·465	48	·42	67	·376
11	·515	30	·464	49	·42	68	·35
12	·515	31	·463	50	·42	69	·35
13	·514	32	·462	51	·42	70	·345
14	·510	33	·46	52	·42	71	·34
15	·504	34	·46	53	·417	72	·34
16	·5	35	·46	54	·407	73	·336
17	·5	36	·46	55	·407	74	·32
18	·496	37	·458	56	·405	75	·281
19	·496	38	·455	57	·4		

The general average of these 75 cases is about $\frac{4}{9}$ D. The shape of the ground, as it widens out above the site of the embankment, makes the average width of the reservoir in all these cases about twice as much as it would be with the straight slopes assumed to meet in the stream at the bottom of the valley, in which case the width of the reservoir is greatest at the embankment; whereas, in the better sites the greatest width is considerably above the embankment, the width there being nearly the average width of the whole reservoir. In the case adduced the area is 3,000 × 450 = 1,350,000sq. ft., and, the greatest depth being 45ft., the capacity of the reservoir is

$$A \times \frac{4}{9} D = 27 \text{ million cubic feet.}$$

SECTION III.

Appurtenances of Reservoirs.

A RESERVOIR for a town supply, as distinguished from a compensation reservoir, sometimes has a bye-channel excavated along the margin of the top-water level from the head of the reservoir to the bye-wash which takes the water from the waste-weir. It is useful for several purposes, even when not absolutely necessary. To begin with, it furnishes a considerable quantity of materials for the bank; it enables the bottom of the reservoir to be laid dry for any purpose; and if a discoloured flood should come down when the reservoir is full, or nearly so, it affords a means of excluding the objectionable water, if made large enough. Where a flood channel is not absolutely necessary, these considerations ought to weigh in favour of a bye-channel being constructed.

At the head of a reservoir it is a good arrangement to have a short bank across the stream for the purpose of arresting sand, stones, alluvium, or what else may be brought down in floods, forming a receptacle known in Lancashire as a lodge, in connection with which the flood-weir and sluices are constructed.

Where it is desirable to separate the clear spring waters of a drainage area from the flood waters, an arrangement, depending for its action on the laws of water in motion, which was first introduced by Mr. Bateman, the eminent engineer, may be applied. It consists of a side channel introduced under and across the line of watercourse, into which the water that arrives at the edge of a weir above it without violent motion drops easily down and is carried by the side channel away from the reservoir, while on the arrival of a flood it overleaps the narrow space down which the purer water drops, and flows over a second weir and away to the flood-water reservoir.

In exposed situations—and such reservoirs as are here described are usually in such situations—the wind acting on the surface of the water surges it against the face of the embankment in such a manner that without some protection the earthwork would be gradually washed away. To prevent this, it is necessary to pitch the face with a harder material, usually stone, to the depth of from 12in. to 18in. Flat-bedded rubble stone makes good work, but where the slope is, as it should be, as flat as 3 to 1, rough rubble spread over it stands well enough where the bank is not exposed to high winds.

It is always desirable to know what quantity of water is flowing out of a reservoir, and in some cases, viz., where mills are compensated by a stated quantity per time measurement, it is necessary that the water should be gauged either by allowing it to flow over a weir, or through an opening below the head of water; but the gauging is perhaps, on the whole, more accurately and satisfactorily made over a weir. To bring the water to a sufficient state of rest before it flows over the weir, it is first received into a basin, out of which it flows quietly.

Sometimes a flood channel is not made, but the water coming into the reservoir during its construction is got rid of through the discharging pipes or culverts, and if more than they will discharge comes down and seems likely to reach and go over the partly-raised bank, temporary shoots are used to convey it over to the outside; and when no flood channel is made, the waste-weir becomes an important feature of safety, or otherwise, according as its length is properly proportioned to the drainage area or not.

No discharge pipes or culverts of practicable dimensions would carry off such a flood as three or four hundred cubic feet of water per second from each 1,000 acres, and yet such a quantity must be expected to come into the reservoir some time or other when it is full. It is plain, therefore, that even with the utmost watchfulness of the reservoir-keeper in opening the sluices on the approach of a flood into a full reservoir, some other and larger means for its escape must be provided, and for that purpose a waste-weir is formed in the solid ground at one end of

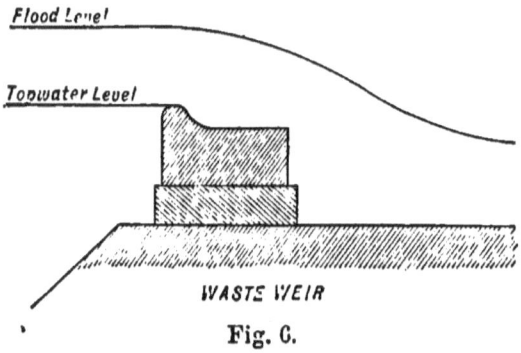

Fig. 6.

the embankment, with a bye-wash to carry the water safely to the stream below. If we consider that with a height of top-bank 6ft. above top-water, or the level of the weir, 4ft. of that space would be little enough to oppose to the wash of water produced by a gale of wind blowing down the reservoir, we shall see that it will be very undesirable, and even unsafe, to allow the water to rise to a greater depth on the weir than about 2ft.

Taking that depth, then, as the greatest to be allowed, and 400 cubic feet per second per 1,000 acres to be the greatest flood, and 100 cubic feet per second to be the greatest quantity that the discharge pipes or culvert would deliver, leaving 300 cubic feet per second to go over the waste-weir, it will be seen that the weir would require to be of a length of 37ft. per 1,000 acres of the drainage area. Here it is assumed that 4 would be the proper co-efficient in the equation $q = c \sqrt{d^3}$, d being in inches, for such a depth over such a weir, which would, from its breadth, if built of ashlar, reduce the co-efficient

THE WASTE-WEIR.

to about that number if the depth were only 1ft.; and, seeing that the greater the depth the more the weir loses its property of a free fall over, or becomes drowned, it may well be supposed that, if it could be experimentally determined for such a depth as 24in., as it has been for depths up to 12in., the co-efficient would be found not to be higher than 4. Using that co-efficient, then, and limiting the depth over the weir to 2ft., it will be seen that 37ft. or 38ft. length of weir is necessary for every 1,000 acres of the drainage area.

This does seem a great length, and it is certainly greater than most waste-weirs possess, but with a shorter weir there might be a risk of the water rising too near the top of the bank, when that is not more than 6ft. above the top-water level. If questions of economy could be allowed to enter as freely into the design of waterworks reservoirs as they do into that of many other works, an engineer might contrast the improbability of an excessive flood occurring at the very time of a full reservoir, and might conclude to save the expense of a long weir, or to allow the water to rise to a greater height upon it than 2ft. or so; but, all things considered, he would be a bold man who would allow the water to rise upon the weir to a height more than half the height of the bank above it, where that height is 6ft., and if the water were allowed to rise to the extent of 3ft. on the weir the length per 1,000 acres would require to be 20ft., for a flood over the weir amounting to 300 cubic feet per second per 1,000 acres of the drainage area.

An instance in which something like this proportion was adopted is the Grimwith compensation reservoir of the Bradford Waterworks. Here the drainage area is about 7,000 acres, and when the late Sir William Cubitt was referred to to fix the length of the waste-weir of that reservoir—being the person agreed upon mutually by the mill-owners and the waterworks proprietors to fix the length of the weir—he prescribed a length of 150ft. This is in strong contrast to the length of another weir in a situation not very dissimilar, and where the drainage area is also nearly 7,000 acres, viz., at Tittesworth, in North Staffordshire, a compensation reservoir belonging to the Staffordshire Potteries Waterworks, where the waste-weir for this very large area was made no more than 60ft. in length, the height of the top of the bank being 6ft. above the level of the weir. The consequence of making the length so short was, that in August of the year 1862 the water rose to a height of 5ft. above the weir; and if only the accidental occurrence of a wind blowing down the reservoir had taken place at the same time, there can be no doubt that a great disaster would there and then have happened.

In the case of the Bradfield Reservoir, a compensation reservoir belonging to the Sheffield Waterworks, which burst in 1864, it had a drainage area of 4,300 acres and a waste-weir of 60ft., in a horseshoe form; and although the bursting of the embankment was due to other causes than the shortness of the waste-weir, there can be no doubt that if the bank had been otherwise soundly constructed upon ground which could not move, its failure would at some time have been imminent from this cause alone.

Although, perhaps, in the greater number of cases the waste-weir is made at one end of the bank, in the solid ground, it has been the practice of some engineers to combine it with the structure necessary to be erected at the inner end of the discharge pipe or culvert, for the purpose of working the valves. This structure consists of an enlarged valve well, of such diameter that the circumference of its top course shall be of sufficient length for a waste-weir. This valve pit affords a ready and convenient access to the inner valve of the discharge pipe, and it admits also of the facility of drawing off the water at different heights, for the water near the top is generally better than that drawn from the bottom, and for this drawing-off purpose valve pits are sometimes built, although they are not made use of as waste-weirs.

SECTION IV.

DISCHARGE OF WATER FROM THE RESERVOIR.

THERE are two distinct principles involved in the practice of drawing off the water from a reservoir; the one is to discharge it through pipes laid so securely and so guardedly as to form a feature of stability equal to that of any other part of the reservoir—as the puddle wall, for instance, or the bank itself—thus rendering unnecessary any provision for inspection or repair, in the same way that one has to trust to the stability of construction of the puddle wall, where inspection is impossible. This mode of discharging the water was frequently adopted some years ago, but of late years not so much, if at all; it consists of one or more lines of cast-iron pipes laid in the solid ground underneath the embankment from the foot of the inner slope to that of the outer one, with valves on them to control the discharge of water. Sometimes these valves are placed at the inner ends of the lines of pipes, and sometimes at the outer ends; the former position being the more secure and the latter the more convenient. Mere convenience, however, in reservoir works, cannot be much considered.

When the valves were placed solely at the outer ends of the discharging pipes they were in duplicate, so that in case of one being out of working order the other might be used. When valves were so placed the mouth of the pipe within the reservoir had sometimes a conical plug suspended over it, so that it might be dropped into the mouth in case of emergency. With large pipes or great heads of water, the pressure on this plug would prevent its being raised again without very great power, if it were not that the main plug contains in its crown a smaller one, which, when it is desired to raise the plug again after use, is first lifted, and, the valve at the outer end of the pipe being closed, the pipe is filled with water and the pressure on the large plug equalised, after which it is easily raised.

There are cases where this precaution of either valve or plug at the inner end of a line of discharge pipe is not taken, but these are cases of the boldest and most self-confident treatment, and verge, indeed, on the indiscreet; and anything even approaching indiscretion in the construction of reservoirs is certainly to be discouraged.

On this principle of drawing off the water the pipes are made unusually strong, the sockets unusually deep, the depth of lead unusually great, and everything, in fact, is done to render the pipe secure under its circumstances. On the other principle the pipe is not made a special feature of stability, but is enclosed in a culvert of masonry of such a size as to admit of inspection and repair of the pipe. It is quite evident that either pipe or culvert should be laid on an unyielding foundation. To admit the possibility of sinking in either case would be absurd. There have been a few cases where pipes have been carelessly laid in this respect, but they have been condemned as bad construction.

If we take first the principle of laying a bare pipe under the bank, it will be seen that an element of weakness presents itself in the liability of the water to follow the line of the pipe between its surface and the material adjoining it. To prevent this it is enclosed in clay puddle, and obstructions to this action of the water are placed at right angles round the pipe. The rims of the sockets themselves form a considerable obstruction to the passage of water, but an additional precaution is sometimes taken to place wider flanges, projecting say, six inches all round the pipe.

The discharging pipes are laid not from the lowest level

of the reservoir, but at a level some feet above the bottom, and it is of advantage to lay another pipe from the bottom itself, which shall have its outlet in the stream below, so that the reservoir may be completely emptied when necessary.

Crossing the puddle trench is the great difficulty of the line of discharge pipe. The puddle will not carry weight, and the pipe has to be made to carry itself and all above it from side to side of the puddle trench, unless it be very wide, when a solid pier of masonry is brought up from the bottom of the trench in the centre, to afford a central bearing to the pipe. The transverse strength of this pipe is capable of estimation as well as any other cast-iron girder, for it becomes a cylindrical cast-iron girder, of such strength as to carry the load due to it, with, of course, in such a situation, a very large margin for contingencies.

Mr. Bateman directed this difficulty to be met, in the case of the Halifax Waterworks, by bringing up a solid ashlar pier of the full width of the puddle trench, dressing down the sides of the trench where the masonry abuts against them, and filling the joint with cement. In the middle of this pier, lengthwise of the trench, an iron plate is brought up, projecting 12in. beyond each face into the puddle, and standing up 18in. above the pipe, the joints between the iron plate and the masonry being filled in with cement, and other precautions were taken in passing the discharge pipe through the iron plate.

There are not wanting on this subject, as on some others, differences of opinion as to the desirability of placing bare pipes under embankments. Culverts of masonry, it is said, are preferable, through which the water may be discharged either openly or through a pipe which can be got at for examination and repairs, and there are some very successful examples of this mode of construction on the Bradford Waterworks, designed and carried out by the late Mr. John W. Leather, of Leeds; as good, perhaps, in design and execution as any in the country.

After the failure of the embankment of the Bradfield Reservoir of Sheffield, where the discharge pipes had been laid on the principle of trusting to the pipe, but where several precautions usually considered necessary had been omitted, it was very much the opinion of Mr. (now Sir) Robert Rawlinson, who, with Mr. Beardmore, investigated the case on the part of the Government, that the bank failed because of defects in the discharging pipes, and Mr. Rawlinson went so far, in some general remarks on reservoirs, as to condemn *in toto* the principle of laying a bare pipe under an embankment, and urged strongly the necessity of enclosing the pipes in culverts of masonry large enough to admit of workmen passing up them.

One ought, however, to consider whether the larger culvert does not invite strains, requiring enormous resistance, which the comparatively small pipe does not, and which is therefore so far preferable.

The joints of a large culvert are no doubt a source of weakness, and it is to this point chiefly, and to that of the possibility of crushing the materials, that attention is drawn when considering whether a large culvert or a small pipe is preferable.

SECTION V.

Approximate Cost of a Storage Reservoir.

To estimate approximately the cost of a reservoir for impounding water by an earthen embankment, the following work would have to be taken into consideration:—

Before depositing any earth, the seat of the embankment would have to be examined, and the top soil and all boggy earth removed outside and reserved for spreading upon the outer slope.

The depth of the puddle trench could not be exactly ascertained before the work is commenced; but it would probably allow a considerable margin for contingencies if the depth were assumed to be, at the deepest part, equal to the height of the bank, and 10ft. at each end.

Inasmuch as the depth to which the puddle trench would have to be excavated would not be known in beginning the work, the sides should be carried down vertically as far as the ground is of uncertain character, and until the ground below can be proved to be strong and suitable for the commencement of the puddling.

In almost every case close-planking the sides of the trench would be necessary. When this is done by driving the planking vertically behind horizontal walings, the depth to which the trench can be carried down of the full width is limited to the length of the runners used, for at the bottom of each set the new set of timbering must be commenced 6in. or 8in. on each side within it, and when the depth of the trench is great, this continual narrowing does not leave sufficient width in the bottom, unless the width at the top be made greater than is required for the thickness of the puddle. It is, therefore, better in these trenches of uncertain depth to lay the planking horizontally.

In going down with the trench, water may be found to come in through fissures in the sides, on one side, or the bottom. As long as it comes through the bottom, the trench would be carried further down, so as to get completely under the water, if possible; but if it come through a fissure in the side only, that fissure would have been passed through, and may be caulked with cotton-waste or tow, or plugged with dry wood.

If the water running into the trench can be stopped in this way, and the bottom is strong and likely to continue so, the clay may be got in and worked for puddle in thin layers all along the trench, doing the puddling more by labour than by water, enough of this only to soak the clay being allowed to run on to it; the bulk of the water being conducted to the sump-hole from which it is to be pumped. The bottom of the trench for this reason should be inclined sufficient to carry off the water quickly. Strong springs of water met with in excavating a puddle trench can only be dealt with by special means; but in nearly every case there will be water naturally soaking out of the adjacent ground when its balanced pressure is relieved, and it is necessary to lay the bottom dry by pumping before the puddling can be commenced.

Nothing could be a worse beginning of the work than to throw a mass of puddled clay into a wet bottom; it could not be united with the sides of the trench by treading, as it can be in the absence of water, and it might well be expected that the water in the ground on the upper side of the puddle would, in that case, rise on its outer side between the puddle and the outer side of the trench; whereas, when once the puddle has been well worked of a stiff consistence, water cannot pass it or act upon it; but if, in the first instance, water is in excess, a way is prepared for a future run which would carry with it small portions of the clay continually.

If the trench cannot be kept free from water while the

puddling is being done, it is necessary to protect it from future contact with water by a facing of concrete and a concrete bed—although, as to the bed, the motion of the water in the bottom carries with it the cement or lime, and leaves little but a mass of loose materials, unless, in the first instance, a drain-pipe be laid under the concrete bed to take the water to the pumping-engine.

In order to estimate approximately beforehand to what depth the puddle trench will require to be excavated, borings should be made along the centre line of the embankment down to and 12ft. into retentive ground; this latter because it would not do to rest the puddle on merely a thin stratum of retentive ground, which might have immediately under it a pervious stratum through which water might pass out under the puddle.

The borings having indicated the surface of the retentive ground, the excavation is to be carried 6ft. into it; and if the borings have truly proved the surface there will then be an assurance of at least 6ft. of sound ground under the puddle. Notwithstanding the borings, however, which are only useful as indicating beforehand the quantity of work to be done, the excavation itself must prove whether it is necessary to carry it still lower than had been approximately estimated.

It is not the bottom at the middle of the bank only that has to be carefully examined—the hill-sides on which the embankment abuts require examination far into them, or at least on that side where the strata are faulty. One of the Manchester reservoirs is said never to have been filled, because, as is supposed, the water, after rising to a certain level, passes round the end of the bank, behind the puddle, through loose strata. It is necessary, therefore, to carry the puddle trench far into the hill-side, where such ground exists at the site of the embankment.

As an example from which to estimate approximately the cost per million cubic feet of storage-room—and that is, perhaps, the most convenient form in which a general estimate can be attempted—one may be taken which would be sufficient to equalise the flow of the streams from an area of 2,000 acres during three consecutive years of least rainfall in a locality where 33in. is the average annual fall of a long series of years; and where, also, the character of the rainfall and the ground are such that 180 would be the proper number of days' storage of the average daily yield of those three years, the site being understood to have the proportions already stated as being approximate to an average of many examples.

In such a case the available depth of a year's rainfall would be 13in., if from 14in. to 15in. be allowed for evaporation and other forms of loss, and from 5in. to 6in. of the 33in. as being in excess of the average of the three years to be reckoned upon. The area from which the water would proceed being 2,000 acres, the yearly amount would be 94,380,000 cubic feet, and the daily average yield 258,575 cubic feet.

To contain 180 days' supply, therefore, of this quantity, the capacity of the reservoir would be 46½ million cubic feet. This would be impounded by an embankment 600ft. long and 60ft. high at the middle of its length. If the top width be made 20ft., the inner slope 3 to 1 and the outer slope 2 to 1, the extreme width of the seat of the bank would be 320ft. nearly (rather less, because the level of the ground at the toe of the inner slope is somewhat higher than at the toe of the outer slope, and because the inner slope is flatter than the outer one); and the area to be cleared of objectionable material, to a depth of, say, 8in. vertically, would be about 12,000 sq. yds.

If the thickness of the puddle wall at the top be made 7ft., and the batter of the two sides 1 in 8, the width of the puddle trench at the lowest part of the ground would be 15ft. nearly, and 7ft. at the ends. The depth in different parts of the trench would vary according to the ground met with, but would be, in general, deepest near the middle, and if at the middle point the depth be assumed to be equal to the height of the bank—60ft. in the present example—and 10ft. at the ends of the bank, a line drawn straight between these lowest points would probably equalise the steps in which, practically, the bottom would be cut, and would represent the average depth. The excavation of the puddle trench would then be, if the sides were carried down vertically, about 8,000 cubic yards. The quantity of puddle below the ground would be the same. Above the ground the puddle wall would be about 6,000 cubic yards, and the total quantity of puddle 14,000 cubic yards.

The embankment would be 80,000 cubic yards including the puddle wall, and deducting that the quantity would be 74,000 cubic yards. The embankment, less the puddle wall, is taken in one, because, although the material consists of three kinds—viz., a toe of rough stone at the foot of each slope, selected material on each side of the puddle wall, and the remainder of the earth, each of which would be estimated at its own price when it had been ascertained from what part of the site each kind of material would be procured—yet in a general estimate the distinction can hardly be made. There would be, however, in this example 8,000 cubic yards of rough stone, 24,000 cubic yards of selected material, and the remainder would be 42,000 cubic yards.

If there are to be two discharges from the same reservoir—viz., one for supply and the other for compensation to the stream—the supply pipe would be laid through the discharge culvert, up to the valve tower at its inner end, whether it be situated in the reservoir, clear of the embankment, or built within the embankment itself. The former is the better plan. The best position of the discharge pipe or culvert, or the line along which it should be laid, demands the most serious consideration in every case, in order to hit the happy medium which combines security with economy.

The level at which it must necessarily be laid being near the bottom of the reservoir, and the puddle trench being almost as necessarily sunk much below that level, the pipe or culvert, when laid near the middle of the bank, must cross the puddle trench at a considerable height above the bottom, and right through the mass of puddle. This is the plan which was formerly adopted, whether the discharge pipe was inclosed in a culvert of masonry or was laid bare in the ground; but as it was necessary to support the pipe or culvert across the trench upon a more solid foundation than the puddle itself afforded, the unequal settlements which took place when the bank had been made to its full height tended to fracture the work.

To obviate this, straight vertical joints completely across the whole structure of the culvert were sometimes provided at the points where the unequal settlements would be likely to occur—slip joints—so that the portion of the culvert between the slip joints might settle bodily and evenly. In the same way the brickwork of railway and other tunnels driven through yielding ground is made with a straight joint at the end of every length, instead of leaving toothing for the next length, which is proper enough, and more sightly when finished, where the ground is in unyielding rock.

This provision of a straight joint, however, in the case of reservoir embankments, has sometimes not had the expected effect, owing to the uncertainty of the direction in which the settlements take place, there being horizontal as well as vertical movements of the earth above the culvert. The culvert, therefore, should not cross the puddle trench at any great height, if at all, above the bottom of the trench. This consideration drives the position a long way from the middle of the bank, as is shown in Fig. 7, in order to lay the culvert in the solid ground.

The bottom of the reservoir, to a height of 14ft. or 15ft. above the lowest point, contains but little water—say in this case half a million cubic feet. It is not necessary, therefore, to lay the discharge pipe lower than is sufficient to draw off the water to this level, say 40ft. below the top-water level, and the lowest sluice should be, say, 3ft.

below this level, to give a head of entrance into the pipe, thus fixing the level of the upper end of the discharge pipe at 43ft. below the top-water level of the reservoir, by which means 46,000,000 cubic feet of water can be drawn off from a full reservoir. If the invert of the culvert be laid 2ft. lower than the pipe, it would form a channel for the compensation water to be discharged into the stream. To determine the size of the culvert, that of the supply-pipe must first be found. The daily quantity

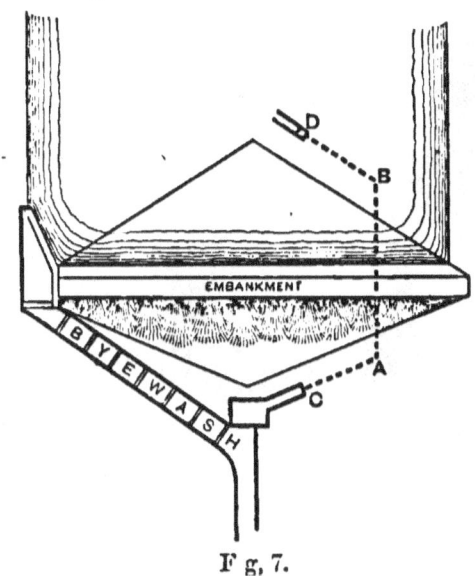

Fig. 7.

of water due to 13in. rainfall on 2,000 acres of ground is, as before found, 258,570 cubic feet.

If one-third be discharged into the stream as compensation, the quantity passing through the pipe would be 172,380 cubic feet per day. If the mean daily velocity through the pipe be made 2 ft. per second, the diameter found from these data would be 15in. In assigning room for the 15in. pipe, which would be laid before the arch of the culvert is turned over it, it should be considered that it might be possible at some future time that it would be necessary to cut out, remove, and replace one of the pipes, in which case more room would be required than is necessary for laying the pipe in the first instance; the width and height, therefore, should be sufficient for this—say 5ft. wide, and 5½ft. high. The best form would be that shown in Fig. 8. The pipe would be supported above the bottom of the culvert, which might be by cast-iron girders built into the walls.

The width of the trench for the culvert would be about 10ft., and it should be made straight as far as it extends under the embankment, and some distance away from the foot of each slope, as shown from A to B in Fig. 7; but excavation may be saved towards the ends by inclining them inwards from the points A and B to C and D respectively.

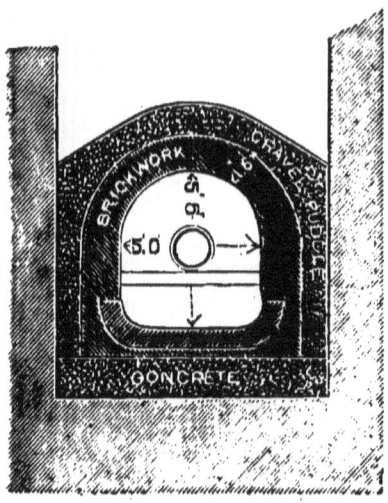

Fig. 8.

In laying the culvert in the trench thus excavated in the solid ground at such a depth below the head of water, every possible care is to be taken to prevent the water finding a way outside it, and for this purpose gravel puddle is a suitable material with which to solidly fill every space not completely occupied by the masonry. It may be thought that concrete is a more suitable material for this purpose, and in any case the invert of the culvert would be laid on a bed of concrete, unless the ground consist of a solid rock; but for filling in the spaces between the back of the masonry and the sides of the trench gravel puddle is a better material, inasmuch as it can be made more compact and watertight in such a situation.

The valve-tower at the upper end of the discharge culvert may be of brickwork 18in. thick, 6ft. diameter inside. The lowest valve being 43ft. below the top-water level; there might be three more above it, at 30ft., 20ft., and 10ft. below that level, so that the water can be drawn off from near the surface at all times. These would all communicate with one descending pipe erected within the valve-tower, from the top of which they would be worked, a light iron footbridge affording access from the top of the bank. Side by side with the lowest valve would be another for drawing off the compensation water. This should be of larger size than the supply-pipe; in this case it might be 2ft. diameter, if circular, but a rectangular form is preferable.

The inside slope of the embankment would be covered with rough stones, the interstices being filled with smaller broken stone, forming a protection to the earth from the wash of water. The area to be so covered would be about 6,000 sq. yds. The outer slope, to be soiled, would be about 4,000 sq. yds., and, including a grass border on each side of the road to be formed on the top of the embankment, the quantity would be 4,800 sq. yds.

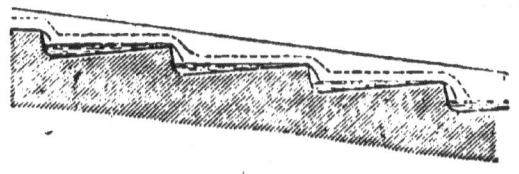

Fig. 9.

A more important part of the work is the waste-weir and bye-wash. For such a drainage area as that assumed the length should not be less than 60ft. The bye-wash may be nearly wholly of concrete, of the form shown in Fig. 9.

By forming the bye-wash in a series of ponds, of not more than 6ft. difference of level between them, the descent of the water, in the heaviest flood, may be sufficiently checked, and it may be gradually let down to the stream below the embankment without injury.

A large body of water may be dealt with safely if it be prevented acquiring a too great velocity. Each of the steps shown in the sketch forms an independent pond into which the water falls from a moderate height. The general inclination is, in the example, 1 in 8, each step 6ft., and the length from step to step nearly 48ft.

There would be, in a flood, a great body of water, but no great velocity in any part of the channel; and, except for the overfall sills, which should be of stone, the whole of the bye-wash may safely be formed of concrete. If the width at the level of each overfall sill be made 30ft., or half the length of the weir, when the water is flowing 2ft. deep over the weir, the corresponding depth over the sills would be 3·17ft.

The land occupied by the whole reservoir would be, on the scale previously referred to, about 50 acres.

The estimate would stand thus:—

	Quantity.	Price.		
		s.	d.	£
Clearing seat of embankment ..	12,000 sq. yᵈ.	0	2	100
Excavation of puddle trench ..	8,000 c. yd.	5	0	2,000
Clay puddle	14,000 ,,	2	6	1,750
Embankment	74,000 ,,	1	6	5,550
Road over embankment	800 sq. yd.	2	0	80
Pitching inner slope	6,000 ,,	2	0	600
Soiling outer slope	4,800 ,,	0	2	40
Excavation for culvert and valve-tower	4,500 c. yd.	4	0	900
Ditto for bye-wash	4,000 ,,	1	0	200
Concrete in foundations	300 ,,	6	0	90
Ditto in bye-wash	2,200 ,,	7	0	770
Brickwork in culvert and valve-tower	700 ,,	24	0	840
Gravel puddle	500 ,,	4	0	100
Stone sills, caps, coping, &c. ..	1,500 c. ft.	2	0	150
Valves, &c.	..	..	..	150
Footbridge	..	..	..	180
Gauge-weir	..	..	..	100
				£13,600
Contingencies				1,400
				15,000
Land and fencing				5,000
				£20,000

This would be at the rate of £435 per million cubic feet of storage-room, for works and land, but as, perhaps, there might be required as much as £2,000 for expenses outside these, the whole cost per million cubic feet would be £478.

SECTION VI.

CONCRETE FOR EMBANKMENTS AND DAMS.

WHERE reservoirs are desired to be made in situations which afford no clay for puddle, and nothing but loose earth and stone for the embankment, the use of concrete may be extended, in forming the bulk of an embankment or dam. Where the range of stone is extensive and massive, and is bedded, it may be more advantageously cut into blocks for setting as proper masonry. The materials of the immediate site, whatever they are, are those only which can be used with economy for the main portion of a reservoir embankment. Where these consist of stone, in any form, they may be used either by way of setting, in such blocks as can be procured, or they may be broken up for concrete, and used for the bulk of the embankment. The material is heavier than earth, and less liable to slip, and for both these reasons it may be used in bulk of less magnitude, with the same degree of resistance.

Concrete may fairly be taken to weigh 120lb. per cubic foot at the least, and much more with some kinds of stone, while 96lb. per cubic foot is a heavy average for earth. The weights of stone of various kinds are approximately as follows, per cubic foot in the solid state, viz.: basalt, 180lb.; granite, 166lb.; mountain limestone, 170lb.; clay slate, 180lb.; trap, 170lb.; sandstone, 144lb. as an average, but some kinds are only 130lb.; chalk, 160lb. When broken up for concrete into pieces of about 4 cubic inches, the space occupied by a cubic foot of solid stone extends to about $1\frac{1}{2}$ cubic foot, more or less as the pieces are moved about amongst each other so that the angles interlock, and the quantity of sand, or fine gravel and sand, with which the interstices may be filled, varies accordingly.

The cementing substance required to combine the mass must be such as will set under water. The beds of blue-lias limestone furnish a hydrate of lime which has this property in a degree sufficient for the purpose; and so, indeed, have some parts of the Wenlock limestone and the gray chalk; but in other situations, where the cementing substance must be brought from a distance, Portland cement will be the most proper material, for the reason that the less the weight to be carried to the spot the better, and that Portland cement will bear a larger proportion of sand.

River-sand, if clean, is better than pit-sand, but it is by no means safe to assume, as is sometimes done, that all river-sand is better than pit-sand, inasmuch as it often contains vegetable and animal fibre in injurious quantity, which cannot be separated from it by any ordinary or economical means. The only means of separating objectionable matter from sand to be used for mortar or concrete is washing it, and river-sand has already had, before it is procured, more washing than can be artificially given to it, and if the objectionable woody fibre, rags, wool, hair, &c., remain in it where it is procurable, they may be considered as being inseparable; and sand containing these, or any of them, in considerable quantity, is unfit for this purpose.

The only way to get rid of the organic matter in river-sand would be to burn it out, but that would be a process too costly to be carried out. Pit-sand, on the contrary, is free from these, but contains too much earthy matter, which requires washing out of it. It is not always possible to do this entirely, with any degree of economy, for in many cases the quantity of clean sand left after the operation of thorough washing would be so small as not to be worth having at the price it would cost; and when neither clean river-sand nor good pit-sand is procurable, crushed sandstone may be used; but it is not a good material for the purpose, inasmuch as that any stone which can be crushed into sand contains much earthy matter. Sand procured in this way is, of course, costly, but even then it is not of good quality, and either of the other kinds is to be preferred to it, when cleansed.

There is another source from which the necessary fine material for concrete may be procured. Clay may be burnt as it is dug out of the ground, at an almost small expense, and if well burnt may be crushed into a fine material resembling sand, which, although not so good a material as clean sand, is preferable to some others, inasmuch as it is absolutely clean; its fault is that it is absorbent, and, if not well burnt, too much so for use. Crushed engine-cinders form another material of similar character, and equally good if procured clean, and consisting of engine-cinders only; house-ashes are, of course, inadmissible under any circumstances.

The immediate purpose for which concrete is intended to be used seems to be not always kept in view in specifying the proportions of its several components.

Where it is used as a foundation to carry weight, or more properly to distribute weight over a larger area of foundation, much sand is to be avoided, inasmuch as it weakens the coherence of the materials as a whole. It is better in this case to use the cementing substance for the purpose of the adherence of the parts of the larger material to each other, and, instead of driving them asunder by interposing sand, to bring them as close together as possible, and let each piece of the larger material be coated with its due proportion of cementing substance. If, after the larger material had been brought as nearly into contact as is practicable, the space of the remaining interstices could be known, they might with advantage be filled; but, as they could not be known, the probability is that, if filled at all, they would be over-filled, and the larger part of the material driven asunder, so that it is probably better to avoid sand altogether.

But where concrete is intended to be used as a wall to prevent the passage of water, the interstices of the material require filling up, and it is important to know what relation of space they bear to the solid material, or to the whole mass, in order that they may be completely filled.

It is understood, and is to be insisted upon as a point of the very greatest importance, that the materials with which concrete is to be made must be clean: no good concrete, of any sort whatever, can be made without attention to this point.

Small angular stones lie closer together than rounded gravel-stones, if means are taken to press them together; but not so without such means. When loosely tipped in a heap, the interstices are larger with angular stones than with gravel, which, without ramming, settles itself to as great a degree of compactness as it is capable of; whereas the other material can be much compacted by ramming. Ramming clean gravel is detrimental rather than useful, inasmuch as it merely displaces the parts of the material without bringing them closer together as a whole.

If the material be neither angular nor much rounded, as beach-shingle, it is of intermediate character in this respect, and may be rammed with some advantage.

If the material were perfect spheres the spaces amongst them could be calculated exactly, thus

The distance, A B, Fig. 10 (next page), is the diameter of a ball; A C being half the diameter, and B C the transverse distance apart of the rows of balls, $= \sqrt{AB^2 - AC^2}$, the longitudinal distance being the diameter of a ball. In the vertical arrangement, Fig. 11, the height $BC = \sqrt{AB^2 - AC^2}$ as before, and the cubic space occupied by each ball is $AB \times BC^2$.

If the balls are 1 in. diameter, the distance $BC = \sqrt{1^2 - \cdot 5^2} = \cdot 866$ in. The vertical height is the same, and the space occupied by one ball is $1 \times (\cdot 866)^2 = \cdot 75$ cubic inch. The solid sphere, 1 in. diameter, is $\cdot 5236$

cubic inch, leaving a space around each ball of ·75 − ·5236 = ·2264 cubic inch, and the ratio of the hollow space to the whole space occupied is ·2264 to ·75, or 30 per cent.

If the balls are 2 in. diameter, the distance apart of the rows of balls is $\sqrt{2^2 - 1^2} = 1·732$ in., both horizontally and vertically, and the space occupied by each ball is $2 \times (1·732)^2 = 6$ cubic inches. The solid sphere is proportional to the cube of its diameter, or to $2^3 = 8$, and is ·5236 × 8 = 4·188 cubic inches, and the ratio of the hollow space to the whole space occupied is 1·812 to 6, or 30 per cent., as before. If the balls are 3 in. diameter, the percentage is the same. The distance apart of the rows is $\sqrt{3^2 - (1·5)^2} = 2·598$ in., and the space occupied

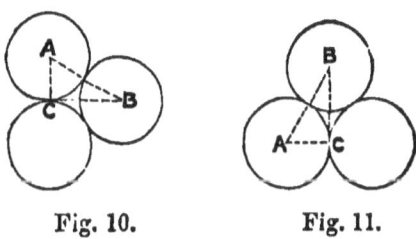

Fig. 10.   Fig. 11.

by each ball is $3 \times (2·598)^2 = 20·25$ cubic inch. The solid sphere is ·5236 × 3^3 = 14·137 cubic inches, leaving a space around each ball of 20·25 − 14·137 = 6·113 cubic inches, and the ratio of the hollow space to the whole space is 6·113 to 20·25, or 30 per cent.

These may be compared with some trials of the proportions of sand and shingle at the Portsmouth Dockyard Extension works, given by Mr. Colson [*] in a paper read at the Institution of Civil Engineers in February, 1881, where 15 samples of shingle of 1 cubic foot each, from various localities—viz., from Langston Harbour, Browndown, and Portsmouth Harbour—were tried, and which showed that it required, on the average of the 15 samples from the above-named localities, 38·4 per cent. of

[*] C. Colson, Esq., M.Inst.C.E.

sand to fill the interstices, or 2·630 parts of shingle to 1 of sand. There was 53·3 per cent. of sand in the material as procured and used in the dock walls and other parts of the work, or 1·875 of shingle to 1 of sand; but in the trials the quantity of sand was reduced to that which was sufficient only to fill the interstices of the shingle, with the above results. At the same time 26 trials were made of the quantity of cement required to fill the interstices of the sand for mortar, which showed a proportion of 36·4 parts of cement to 100 of sand, or 2·715 of sand to 1 of cement.

Mr. Deacon,* of Liverpool, made some trials of concrete used in the foundation of roadways, given in a paper read by him at the Institution in April, 1879. The concrete consisted of 8 parts of broken stone, 6 parts of gravel, and 1 of cement, making a mass, when mixed and beaten together, of 11 parts of stone and gravel to 1 of set cement; from which it would seem, the cement being included in the 11 parts, that it required 3 parts of gravel (containing half a part of cement) to fill the interstices of the broken stone, there being in the produced mass 3 parts of gravel more than was sufficient to fill the interstices. The percentage of space of the interstices of the broken stone to the whole mass was thus 3 to 11, or 27·27 per cent.

When the quantity of sand and cement is considerably in excess of that required to fill the interstices of the larger material, as in this latter case and in the walls of the Portsmouth Dockyard extension, it takes the form of a matrix in which are embedded the larger pieces of material, and thus becomes or resembles rubble masonry, such as is found in the old castle-walls and Roman buildings, in which pieces of rough stone are embedded in a coarse mortar. This is the form which concrete walls should take which are intended to retain water in a reservoir. Concrete has not hitherto been used to form a reservoir embankment entirely, but during the last few years it has been used to protect the puddle walls of earthen embankments.

* G. F. Deacon, Esq., M.Inst.C.E.

SECTION VII.

Stream Gauges.

The watershed line along the tops of hills sheds the rainfall in opposite directions into different valleys—the basin, when taken as a whole, in each case, from source to sea; or the catchment area, or drainage area, or watershed area, when a part only is dealt with.

Of the rain falling upon this watershed area and running off in streams, some is lost by evaporation in the atmosphere and absorption into the ground before it arrives at the point down to which the area is calculated, and the quantity lost can only be ascertained by gauging the flow of the streams and comparing the quantity so found with the whole quantity due to the area and depth of rainfall. The quantity lost varies with the declivity and character of the ground.

After a sufficient number of gaugings of the streams has been made, the practical quantity which can be depended upon may be ascertained. Such gauges, being of a temporary nature, may be made with planks, the joints being grooved and tongued, and the whole bolted together and stiffened with battens. A stream should be gauged daily at least once, but that is not sufficiently often for accuracy, for within twenty-four hours the depth of water flowing over the gauge may vary greatly. Twice a day, or three times, is not too often—say at 6 A.M., 1 P.M., and 8 P.M.—and even these frequent observations need to be supplemented by notes of the beginning and end of rains sufficiently heavy to influence the stream of water; but these are not so necessary if the twenty-four hours be divided into three equal periods of time for the observations of the depth.

In any gauge of this sort, whether temporary or permanent, it is necessary to dam up the stream of water and form a considerable pond in which the water may expand itself and flow out quietly over the gauge, and from as nearly still water as possible. If this precaution be not taken the velocity of the water before arriving at the gauge would have to be measured at the time of every observation, and this would admit numerous possibilities of error.

Hydraulicians have measured the actual quantities of water flowing over weirs and notches in thin plates and plank-gauges, by receiving into a large tank below the gauge all the water flowing over it in a given time at various depths, every condition of the flow of water and the construction of the gauge being observed at the same time—such as the width of the gauge with reference to the width of the pond; the thickness of the lip or sill over which the water flows; whether the edge be horizontal, crosswise of the gauge, or bevelled off, and if bevelled, which direction, whether outside or inside; and the same with respect to the ends of the openings; the velocity of the current (if any) before arriving at the gauge; and whether the depth recorded be that from the surface of the water in the pond or that directly over the sill of the weir or in the notch; because this latter is always considerably less than the former, and requires the application of a different rule, and not one of these particulars can be omitted to be recorded if the gaugings are to have the value which the labour bestowed upon them would otherwise give them.

The depths have been mostly measured from still water, and the results have been formulated accordingly; but it is not sufficient to assume that this has always been the depth measured, unless it has been expressly stated so. Some of the experiments have been made over thin plates of copper or iron, some over thicker plates of iron, some over boards of various thicknesses, and as these and other particulars have varied, the actual quantities passing into the outer tanks have varied for the same depths; and in adopting a thickness of the lip of the gauge, regard should be had to the conditions of the experiments relied upon, so as to place the gauge under similar conditions to those of one set or another of these.

Almost any set of experiments may be relied upon if the conditions under which they were made be duly observed; but if they be taken together indiscriminately they form nothing better than a muddle of facts from which nothing certain can be determined. For this reason, in fixing a temporary plank-gauge, for instance, where the water can be dammed up into a pond large enough to prevent it reaching the gauge with any distinct current, 2in. is a better thickness than 3in., because a greater number of accurate experiments has been made over planks 2in. thick than over 3in. planks.

And with reference to other experiments, the same precaution should be taken. Mr. Blackwell's* experiments were made over various thicknesses of lip and widths of sill, thin plates, 2in. planks, and wide-crested weirs, and over various lengths of gauge with the same depth of water, —3ft., 6ft. and 10ft. Those over 2in. planks were 120 or more in number, from 1in. to 14in. in depth.

Those experiments of Mr. Blackwell, above referred to, were made on the Kennet and Avon Canal, and were made from a still head of water; but it will often happen in gauging streams that a pond large enough to effect this cannot be made, and in that case a set of experiments— also over a gauge 2in. thick, of iron—made at one of the reservoirs of the Bristol Waterworks, more resemble the conditions of temporary stream gauges, the pond being not so large, and the water arriving at the gauge with some current, though not much. The co-efficients found from these experiments were therefore greater than those found from the experiments first referred to.

* The late T. E. Blackwell, M.Inst.C.E.

This latter set of experiments was made by Mr. Blackwell in conjunction with the late Mr. James Simpson, the engineer of the Bristol Waterworks, who furnished to him a set of observations he had previously made on the same weir, which is the weir of the gauge-basin of the Chew Magna Reservoir, and is 10ft. long. The experiments were about 60 in number, from 1in. to 9in. in depth.

If it is intended to rely upon those experiments which have been made over thin plates, the number of which is, perhaps, greater than has been made with planks of any thickness, although most of those over thin plates have been on a smaller scale, a strip of sheet-iron may be screwed on, on the upper side of a plank-gauge; but in that case it should project clearly beyond the edge of the plank, so that the edge of the plank may have no influence upon the flow of water, and when a strip of iron is thus attached on the upper side of the opening, the edges of the plank should be bevelled outwards, away from the iron plate.

With reference to the dimensions of a gauge, regard must be had to the area of the drainage ground and to the probable variation in the rate of flow of the water. The gauge must allow floods to pass through it or over it, and at the same time it must be small enough in width to gauge the dry-weather flow. The one quantity may be a thousand times more than the other.

The variations being so great, it is necessary to confine the smaller quantities to a less width than that of the full gauge, by making a notch in the gauge of such limited width as will afford a measurable depth, and one which will accord nearly with some of the depths which have been experimented upon.

The width of the notch, as well as that of the whole gauge, should be apportioned in some degree to the area of ground from which the water flows. To take, for an instance, 300 acres, there may flow from it as small a quantity as 10 cubic feet per minute, or less, and a flood may amount to 9,000 or 10,000 cubic feet per minute. If the notch be made 2ft. wide, 10 cubic feet per minute would pass through it with a depth of 1in., or thereabouts; and if the depth of the notch be 6in., the quantity would be about 140 cubic feet per minute when running full. A considerable rainfall would then have taken place. Greater quantities of water than this would extend over the whole width of the gauge.

To avoid errors, it is well not to subdivide the width, and form more than one notch. If the full width be made 6ft., the gauge would pass 3,000 cubic feet per minute when running 1ft. 9in. deep, and this would be deep enough for the full depth of the gauge, in consideration of its strength.

Floods greater than 10 cubic feet per minute per acre of the drainage ground would pass over the top of the gauge, and could not, perhaps, be accurately measured; but the gauge ought to be so set that a proper record of the depth can be made, and some tolerably approximate estimate made of even the greatest flood.

To resist floods, the gauge must be secured in masonry walls at the ends, the ends of the plank-gauge being built into them, or by piling, according to the nature of the ground. The first thing to be done in fixing the gauge is to divert the water from its channel, if possible; but if not, to divert it from each half alternately, by a clay dam, and to pump out the water and excavate a trench across the bed of the stream and well into the banks on each side, of a width of about 3ft. and about the same depth, to be filled with puddled clay; and when a depth of 2ft. of puddle has been formed in the bottom, the gauge should be set level upon it, and the filling of the trench on the upper side of the gauge completed up to the bed of the stream.

On the lower side the puddle will be covered with the pitching-stones of the apron, the top of which should be at least 6in. below the notch. If the pitching-stones are 18in. deep, this makes the depth of the gauge below the notch 2ft. If the depth of the notch be 6in., and the top of the gauge 1ft. 9in. above the top of it, the whole depth of the gauge at the ends will be 4ft. 3in. The length of the gauge would be 20ft., the faces of the

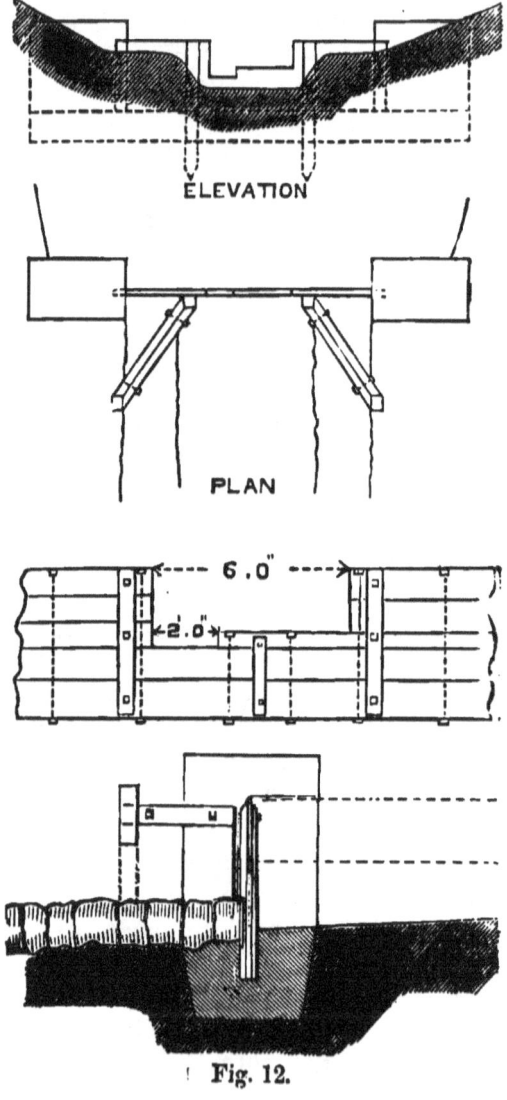

Fig. 12.

abutment walls between which it is fixed being 18ft. apart, and each end of the gauge should be built 1ft. into the walls. Such a gauge might be made of 3in. planks, the edges being reduced to 2in., unless thin iron plates be attached. It would require to be strongly battened, and it should be supported by piles, 12 by 12, in the positions shown in diagram on the preceding page, strutted from other piles in the banks of the stream. To allow the greatest flood to pass over the top of the gauge with a depth not exceeding 15in., the total length between the walls would require to be 18ft., and the walls should be carried up to that height above the top of the gauge, and as far into the banks as to prevent the passage of water round them.

It must not be overlooked that the depth of the puddle-trench named, 3ft. below the bed of the stream, is but little, considering the height of the water above it in a flood, which might be $4\frac{1}{2}$ft. or 5ft., and it is possible that the water might blow a hole through the ground under the puddle. It is a contingency to be considered, but it is not a thing which would probably occur, and the apprehension of it need not prevent a gauge being fixed in the manner described. At the same time, if circumstances are favourable, it might be better to carry the puddle-trench deeper.

In the case of a very loose bottom of sand and gravel—but not in a compacted gravelly-clay bottom—a row of sheet-piling would be preferable to a puddle-trench, for which 3in. deals would be sufficient, and if the bottom is silt, wholly or in great part, piling becomes a necessity, as in such ground the water would be very likely to flow out under the puddle. It is important that the gauge not only be set level from end to end, but that it remain so, and a gauge of this length should be supported by two intermediate bearing piles, or large stakes, driven into the bed of the stream, 4in. to 6in. square, or 5in. to 7in. diameter; and these intermediate supports are equally necessary, whatever the ground may be.

The dimensions of the gauge shown in the diagrams are suitable for an area of 300 acres or thereabouts, and for larger areas it will be preferable to gauge the quantity of water at two or more different points of the area if more than one stream proceed from it, rather than, for a temporary gauge, to place one to command the whole area; for a plank-gauge cannot well be made longer than about 20ft., or wider than 4ft. or 5ft.

Below the bottom of the notch 2ft. is little enough, while 2ft. above the top of it is high enough, in consideration of its strength to resist a flood, and, indeed, this it would not do, with these dimensions, unless firmly supported laterally near the middle, either by struts from the banks or by raking struts from piles driven into the bed of the stream, or by raking walls built in the same positions. The planking, indeed, is to be looked upon as being merely a watertight shield, without lateral strength.

The width of the pond immediately above the gauge should be considerably wider than the total length of the gauge—in this case not less than 30ft.—and the depth of water should be measured at a distance of 6ft. above the gauge, for which purpose a stake should be driven as far from the edge of the pond as will allow the divisions of the scale to be distinctly read, and on the stake a scale divided into hundredths of a foot, or into tenths of an inch, should be fixed with its zero level with the bottom of the notch.

The depth of the water passing over the gauge-weir affects the quantity in two respects; first, as being directly proportionate to the quantity, as the width is, the two dimensions forming the sectional area calculated upon; and, secondly, it affects the quantity as being a measure of the velocity with which the water passes over, this being proportionate to the square root of the depth.

Dividing the cubic quantities measured in a tank, which have passed over the gauge-weir at known depths, by all the dimensions before stated, numbers result which are the constant multipliers required to convert the general formula to practical use.

If c = the constant multiplier.
w = the width of the gauge in feet.
d = the depth of water in inches.
Q = the quantity in cubic feet per minute.
$$Q = c\,w\,d\,\sqrt{d} = c\,w\,\sqrt{d^3}$$

The values of c, derived from Mr. Blackwell's experiments before mentioned, made on the Kennet and Avon Canal, are as follow for 2in.-plank weirs, and for iron plates $\tfrac{1}{16}$in. thick.

Depths.	2in. planks.	Thin plates.
1in.	3·50	5·72
2in.	4·25	5·70
3in.	4·44	4·91
4in.	4·44	4·90
5in.	4·62	4·83
6in.	4·57	4·57
7in.	4·61	
8in.	4·48	4·48
9in.	4·44	4·08

These are the means of the values obtained for the several widths of 3ft., 6t., and 10ft., in the case of 2in. planks, and of 3ft. and 10ft. in that of thin plates. It is not, as a rule, advisable to take averages upon which to base future calculations—that is to say, averages of results, although each result should be the average of a number of experimental observations of the same thing, to insure accuracy. But in this case it may be observed, on an examination of the experiments, that there is less reason to refrain from adopting an average of lengths than of depths; the depths, therefore, and their corresponding co-efficients, are given in detail, while, in respect of width of opening, the means of the values of the several co-efficients are set down.

Mr. Beardmore, a very good authority, gave 5·1 for thin plates. Mr. Francis, of Lowell, U.S., gives what is equivalent to 4·81 for a plate 1in. thick, but rounded off on both arrises, so that there is only ¼in. flat on the top.

But for temporary stream-gauges it is hardly possible to bring the water to a still head, such as the values of c

above stated demand; and probably those derived from the experiments at Chew Magna, before referred to, would give results nearer the truth in such cases. They are approximately as follow, for an iron plate 2in. thick, flat on the top, the width of the opening being 10ft.:—

Depths.	Values of c.
1in. to 2in.	5·0
2in. to 2¼in.	5·1
2¼in. to 2½in.	5·2
2½in. to 2¾in.	5·3
2¾in.	5·4
3in.	5·5
3in. to 4in.	5·6
4in. to 5in.	5·7
5in. to 6in.	5·8
6in. to 7in.	5·8
7in. to 8in.	5·7
8in. to 9in.	5·6

It may be observed in general that the fundamental law of water falling over a weir is the law of gravity as it affects a heavy body falling freely, which is, if we put $g =$ the force of gravity $= 32$, that the velocity in feet per second acquired in falling through the height h in feet is proportional to $\sqrt{2gh} = \sqrt{64h} = 8\sqrt{h}$, neglecting a small decimal of ·02 or ·03.

The height from which the lowermost particles of a sheet of water going over a weir fall, is the height from the surface of the water above the weir, where it is comparatively still, above the crest of the weir; and as the velocity of the particles at every point in the height decreases upwards as the square root of the depth below the surface, the mean velocity of all the particles of the sheet of water is necessarily two-thirds of that due to the lowermost, the velocity due to which is $8\sqrt{h}$, and the mean velocity of all its particles from bottom to top is ⅔ of $8\sqrt{h} = 5\frac{1}{3}\sqrt{h}$, by the abstract law.

But when the weir or notch through which the water runs is much less than the width of the pond above it, the sudden and various changes of direction which the particles follow offer a resistance to their free flow, so that practically

DEPTH MEASURED FROM STILL WATER.

not more than about five-eighths of the calculated abstract quantity passes over a weir—on an average of all circumstances—and if the different circumstances of weirs could be neglected, the quantity in cubic feet per second would be $\frac{5}{8}$ of $5\frac{1}{3} = 3\frac{1}{3} \sqrt{h^3}$ per foot in length of the weir.

If the depth be measured in inches, the constant $3\frac{1}{3}$ should be divided by the square root of the cube of 12, $= 41\cdot56$; but then, when depths are not great, it becomes better to take the quantity in cubic feet per minute instead of per second, and in this case $3\frac{1}{3} \times 60 = 200$, and $\dfrac{200}{41\cdot56} = 4\cdot81$, the constant multiplier for these measurements. It is less than is sometimes adopted, but is as much as it is safe to adopt without taking into account all those particulars of a weir which bear upon the obstructions or facilities to the flow of water over it.

This constant 4·81 is applicable to what may be called the normal condition of a weir, in which the water falls over the weir or notch from a still head, and when the height h is the whole height from the sill of the weir to the surface of still water above it—say four or five feet above it—and when there is no velocity of approach.

When the water arrives within a few feet of the weir in a stream with some sensible velocity it must be taken into account; it is equivalent to an increase of head forcing the water over the weir, and this head is one sixty-fourth part of the square of the velocity in feet per second with which the water flows, at its surface, towards the weir.

The condition which most influences the quantity of water passing over a weir, when calculated from the whole height (h), is the thickness of the lip or sill over which the water flows. It is only when the thickness is reduced to a minimum, and eliminated altogether from calculation, that even an approximation to the true quantity of water flowing over a weir can be arrived at by the application of any constant, such as 4·8, or perhaps more commonly 5 or 5·1; and indeed for accuracy every weir requires a careful study of its circumstances.

The influence of thickness of lip or sill, over which the water flows, is shown clearly by some trials made by the author, the results of which were communicated to the Institution of Civil Engineers, and published in the Minutes of Proceedings, vol. xc. p. 305. One set of experiments was made over a ½in.-plate weir, another set over a 3in.-plank weir. The whole depth (h) varied from 1in. to 7in. The diminution of depth on the outer edge is produced by the retardation of the flow over the sill of the weir, the thicker sill retarding the flow of water more than the thinner one, and a thin plate, say $\frac{1}{18}$in. thick, would, if tried in a similar way, probably show still less retardation than the ½in. plate. Some few experiments were made by the author with sills 6in. and 12in. thick, and these showed much greater reductions of depth than those with the 3in. plank, but they were not made with sufficient accuracy to warrant their insertion along with the others in the paper communicated to the Institution. Such as they were, however, they showed a greater diminution of depth than the weirs of less thickness. The fact, however, which is the most evident upon an examination of every experiment, with whatever thickness of sill, is that there is no simple ratio between the whole depth (h) and the depth on the outer edge (d), but that $\dfrac{d}{h}$ varies with every depth, in such manner that with the ½in. plate

$$d = \cdot 72\ h + \cdot 025\ h^2$$

and with the 3in. plank

$$d = \cdot 33\ h + \cdot 05\ h^2$$

These are in some measure confirmed by a few experiments on the outer edge of 2in. plank-weirs, made by the late Mr. T. E. Blackwell. These show that the curve which would most nearly traverse the points determined by the experiments is one of which the following would be the equation, viz., $\qquad d = \cdot 60\ h + \cdot 025\ h^2.$

DEPTH ON THE OUTER EDGE OF A WEIR.

The following diagram, Fig. 13, shows all three curves. The scale is one-third of the full size. By plotting the dimensions to the full size, and drawing the curve through the points determined by the rules just given, the depth on the outer edge resulting from every whole depth may be accurately measured.

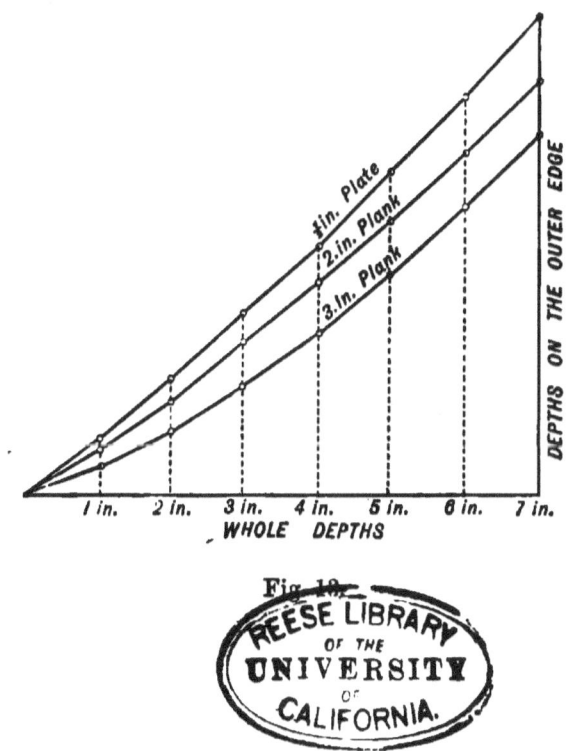

Fig. 13.

SECTION VIII.

Rainfall.

THE chief source of the rainfall of this country is the Atlantic Ocean. The heat rays of the sun convert water into vapour; and as the vapour of water is lighter than air, being about five-eighths of the weight of air, bulk for bulk at the same temperature and pressure, it rises into the atmosphere and collects together in clouds, which are blown about by the wind, and chiefly by the prevailing westerly and south-westerly winds, moving laterally and rising up against the hill-sides into a colder region, where it contracts and unites in rain drops, for as it is heat which changes the particles of water on the surface of the ocean into vapour, so it is the loss of heat which changes the vapour into water again. The vapour itself is invisible; what is seen as clouds is partially condensed into water, but not yet into globules heavy enough to fall through the resisting medium of air.

Rain is precipitated in large quantities on the high ground of the chain of hills which range in a generally north and south direction through England, which shed the water east and west—eastwards into the great basins of the Thames, the Great Ouse, the Trent, the Humber, the Yorkshire Ouse, the Tees, the Tyne, the Tweed, the Forth and the Tay, and westwards into those of the Clyde, the Eden, the Ribble, the Mersey, the Dee, and the Severn, with other (minor) river-basins; while on the south there are the Avon, the Parret, the Exe, and the Tamar. This dividing ridge begins in the neighbourhood of Devizes, and upon or near it are situated Wootton Basset, Thames

MAIN RANGE OF HILLS. 65

Head, Cheltenham, Daventry, Naseby, Coventry, Birmingham, Wolverhampton, Newcastle-under-Lyme, Buxton, Chapel-en-le-Frith, Saddleworth, Rochdale, Colne, Settle, Ribble Head, Swaledale Head, Tindale Tarn, Tyne Head, Tweed Head, and Clyde Head; and so into the rocky formations of Scotland.

These hills intercept the clouds of vapour and condense a portion of them which is not able to rise over their tops, precipitating a large amount of rain, the fall being generally from 30in. to 40in. in a year, and as much as 70in. or 80in. in particular situations; while other portions of vapour follow the western slopes until they come to a gap or break in the range of hills, through which they pour and are condensed on the eastern side, giving at these particular places as much rain as on the western side.

If the rainfall on the western coast be compared with that further inland, it will be seen to increase as these hills are approached. Mr. Bateman drew attention to this in his evidence before the Water Supply Commission, to the effect that if in any particular year the rain on the westerly coast, say, at Liverpool, is 36in., and going right across the country to the east coast, it increases as the range of hills is approached, where it is at the foot, say, 40in., on the summit of the hills it is from 50in. to 60in.; further east it is 30in., and further east still it is 20in. In an investigation he had made across England from Liverpool over the Manchester Waterworks drainage-ground and over that of Halifax, it was observed that the winds impinge upon the westerly slope of the mountains which form the eastern side of the Rivington (Liverpool) Waterworks drainage-ground, where in a certain year the rainfall was 48½in.

Over the ridge, in the first trough to the east, are the Bolton Waterworks, the trough running pretty nearly north and south; and at Belmont, in the Bolton Waterworks district, the rain was 53in. The next trough is the

F

valley from which the Blackburn Waterworks are supplied, and there it was 42in.; and over the mountains right down in the east it was 30in.

The Manchester Waterworks are formed in a mountain range, the Pennine chain of hills, lying between Manchester and Sheffield, commonly called the backbone of England. Rain at Manchester in 1859 was 38in., in round numbers. Going eastward, the rain at the Denton Reservoir, which is 300ft. above the sea, was 34in; at Newton it was 35in., and a little further east it was 33in. All these places are upon the plain, or nearly the plain. Then at the foot of the hills the depth of rain was $46\tfrac{1}{4}$in. Higher up at the head of the valley, on the west side of the hills, it was $53\tfrac{1}{3}$in., then 57·64in.; a little further towards the head, on the east side, just over the summit, it was $58\tfrac{1}{2}$in.

At the reservoir of the Sheffield Waterworks, which is on the hills more to the east, it was 46in., at Penistone 39in., and at Sheffield 25in., showing an increase as the hills are ascended, and, over them, rapidly diminishes towards the east.

On the line of the Rochdale Canal, which also crosses the backbone of England, in the same way, in another year, the rainfall at Rochdale, which is at the foot of the hills, was 39·7in. At Whiteholme, on the top of the hills, 66·7in.; at Blackstone Edge tollbar, 67·5in.; and, on the east side, 66·6in.; but these last three places are almost entirely upon the summits of hills not more than 1,200ft. high. Then at Sowerby Bridge, which is at the easterly foot, though comparatively in the hills, it was 32·27in., and at Halifax less than that.

It was observed by Dr. Miller, in the Lake district, that the rainfall increases up to a height of about 2,000ft. above the sea, but decreases on mountains of higher elevation.

If the elevation of the country lies within the region of the rain-clouds, which may be said to extend to about

3,000 or 4,000ft., the greatest portion of deposit within that range takes place at from 700 to 2,000, or 2,300ft. If the greater portion of the gathering-ground lies within that zone, setting aside local circumstances, that will be the elevation which will give the greatest quantity of rain.

Supposing a continuous ridge exceeding 2,000ft. high to range north and south, there would be comparatively little rain on the east side; it would stop nearly all the rain-clouds from the west, and the water would be precipitated on the westerly slopes; but, where there is a range of mountains, the summits of which measure upwards of 3,000ft. with depressions in the ridge, the largest amount of rain falls in the valleys to the east of those depressions. These westerly and south-westerly winds prevail over the greater part of England and Wales; but in the north of England, in Northumberland, a westerly wind is a dry wind, and a wet wind on the coast of Northumberland is a dry wind in the west.

Mr. Symons began in the year 1858 to collect and arrange the observations which had then been made of the depths of rain fallen in previous years, and to extend the observations, and since 1860 has published them annually through Mr. Stanford, of Charing Cross. There are more than two thousand places in this country where the rainfall is observed daily and sent to Mr. Symons at Camden Square. The observations are examined, collated, and put into form to be useful both in the localities whence they originate and for comparison with the observations of other localities, the whole forming a yearly record of great value.

The rainfall map of the 6th Report of the Rivers Commission, 1874, shows the average annual fall at various places in most of the river-basins. Taking first those on the western coast, which have the heads of the valleys towards the east, the rainfall is seen to increase upwards from the coast as the watersheds are approached, thus:—

		Inches.
Mersey basin:	Liverpool	35
	Manchester	36
	Marple	36
	Chapel-en-le-Frith	40
Ribble basin:	Preston	39
	Settle	50
Lune basin:	Lancaster	44
	Garsdale	52
Lake district:	Whitehaven	52
	Kendal	61

Taking next, from the same records of rainfall, those river basins in which the heads of the valleys are towards the west, it is seen that in general, but not without exceptions, the depth decreases from the watersheds towards the middle and lower portions of the basins, thus:—

		Inches.
Thames basin:	Cirencester	31
	Oxford	25
	Greenwich	24
Severn basin:	Hengoed	36
	Shrewsbury	27
	Worcester	28
Wye basin:	Rhayader	46
	Hereford	30
	Ross	27
Trent basin:	Birmingham	31
	Derby	26
	Retford	23
Calder basin:	Halifax	31
	Pontefract	25
Aire basin:	Head of the valley	35
	Leeds	23
Yorks. Ouse basin:	Richmond	31
	Thirsk	24
	York	23
Tyne basin:	Alendale	50
	Tynemouth	27

On the "Map of the Rivers and Catchment Basins" of the Ordnance Survey Office a series of rainfall depths are given which, although not the average rainfalls, are comparable one with another, and, for this purpose, are perhaps better than averages would be. These show the following reductions of depth of rainfall from west to east:—

COMPARATIVE RAINFALLS.

		Inches.
Thames basin:	Head of the valley	31·8
	Oxford	27·1
	Reading	24·8
	Windsor	23·6
	Kingston	24·2
	Croydon	24·2
	Greenwich	28·5
	Sheerness	22·5
Severn basin:	Head of the valley	68·4
	Shrewsbury	23·5
	Worcester	26·0
	Cheltenham	25·7
	Gloucester	21·8
Wye basin:	Rhayader	41·0
	Hay	32·7
	Hereford	28·3
Trent basin:	Birmingham	31·0
	Leicester	27·6
	Loughborough	26·3
Derbys. Wye and Derwent basin:	Head of the valley	65·1
	Cromford	37·3
	Belper	29·9
	Derby	29·4
Don basin:	Head of the valley	54·8
	Barnsley	27·4
	Sheffield	32·5
	Doncaster	33·1
Calder basin		49·7, 31·1, and 27·2
Aire basin		39·0, 27·7, ,, 24·2
Wharfe basin		54·7, 30·9, ,, 25·2
Yorks. Ouse basin		27·0, 25·0, ,, 25·5
Wear basin		31·3, 25·8, ,, 24·2
Tyne basin		45·5, 28·5, ,, 24·2

It is only by a patient attention to and collection of observations on the rainfall in various parts of the country that data can be arrived at upon which to base calculations of what quantity of water may be expected to be derived from any particular source in any given locality. All atmospheric phenomena are most intricate in their relations to and effect upon the land, and nothing but the most patient study of observed facts, extending over long periods of time, can determine anything worthy of being relied upon in practice.

The rain-gauge is the instrument by which the quantity of rain falling to the earth is measured. There are various forms of it, from the rude quart bottle with a funnel inserted in the neck, to the most delicate instrument of the scientific professor. Some are made to carry a float in the body of the gauge, to which is attached a graduated rod projecting from the mouth of the funnel, which by its rise indicates the depth of rain fallen. Others have a graduated glass tube attached to the body of the gauge, with a communication between them at the bottom, the water rising to the same level in both.

Others, again, are made with a loose funnel inserted into a vessel which is to be emptied into a separate graduated glass tube when it is desired to know the depth of rain fallen. There are also various modifications of each of these kinds of gauges by the various makers of such instruments. The best size of the mouth of the funnel is not very well determined, but a common size is 8in. diameter; others, again, are 10in. diameter, but Mr. Symons, who has under his control a great number of rain-gauges in the kingdom, and who has paid a great deal of attention to the niceties of measurement, said that a funnel of 5in. diameter is as good as any other, and sufficiently accurate.

The height at which a rain-gauge is fixed above the ground is important to be attended to. It has long been known, but apparently not generally known, that the nearer the ground the gauge is set, the more rain it registers, but it was not until Mr. Symons's comparative experiments had been made, that anything like a definite conclusion on this point had been arrived at. Gauges were fixed at all heights from the surface of the ground to a height of 20ft.

At the ground-level itself the cause of the greatest register of quantity was proved to be because of the rebounding into it of rain-drops falling on the surrounding soil or grass. When, however, a trench was dug round the gauge so as to prevent this, the register ceased to show excess. On the whole, Mr. Symons recommends the mouth of the rain-gauge to be not less than 6in. or more than 12in. above the ground, except when a greater elevation is necessary to obtain a proper exposure, for this is important.

Gauges sheltered by buildings, or trees, or shrubs, or tall flowers, are not in the best situations. The distance from any building should be at the very least as great as its height. The gauge must be fixed perfectly level. If it be impossible to fix the gauge unless near some buildings, those are to be preferred which stand north-west, north, and north-east of the gauge; those standing south and south-east will be in the next least objectionable quarter, and those standing south-west of the gauge are in the most objectionable positions. Observations should be made daily, and at the same hour of the day.

In snow, that which is caught in the funnel should be taken out and melted and measured as rain. As a check upon this, select a place where the snow is of a fair depth, not drifted, invert the funnel and take up a funnel full, or whatever can be taken up in this way, and melt it. As a second check, measure the average depth of snow and take a twelfth of it as the equivalent depth of water. Strike an average of the three processes and enter that as the depth of rain. But it is to be observed that in respect of the first of these methods the snow is liable to be blown out of the funnel if the weather be at all windy, and allowance, therefore, should be made for this. Snow, measured in depth within a few hours after it has ceased to fall, has often been found to yield, by melting, a depth of water equal to one-twelfth of its depth, not unfrequently one-tenth, and sometimes one-eighth; it depends upon the state of the atmosphere at the time of the fall.

SECTION IX.

Areas of River-basins.

The areas of the river basins of England and Wales, prepared by the Ordnance Department and published by Mr. Stanford, of Charing Cross, may be divided into sections, as follows, beginning at the Land's End, and proceeding along the south-east and north-east coasts, and along the north-west and west coasts back to Land's End.

From Land's End to Fowey, a distance of 85 miles along the coast—not following every indent of the sea, but a general line—12 rivers or streams have basins of 47, 40, 29, 33, 10, 33, 12, 40, 66, 50, and 80 square miles respectively, and the river Fowey itself of 120 square miles.

Proceeding in like manner, a further distance of 88 miles from Fowey to Dawlish, there are 12 rivers having basins of the following areas:—

Name.	Area.	Name.	Area.
—	71	Yealme	36
Lynher	100	Erme	43
Tamar	385	Aune	54
Tavy	85	—	73
Plymouth Leat	23	Dart	200
Plym	59	Teign	203

From Dawlish to St. Alban's Head, a distance of 90 miles, there are nine rivers, the areas of the basins of which are, in square miles, as follows:—

Name.	Area.	Name.	Area.
—	11	Char	39
Exe	584	Brit	52
Otter	82	Bredy	21
—	21	(Weymouth)	87
Axe	165		

SOUTH AND EAST. 73

From St. Alban's Head to Littlehampton, 95 miles, there are 13 rivers—

Name.	Area.	Name.	Area.
Frome	187	Itchen	231
Piddle	119	Hamble	35
Stour	459	——	85
Avon	673	(Portsmouth)	235
Lymington	91	——	26
Beaulieu	52	Arun	370
Test	477		

The Isle of Wight is not included; it has a coast-line of about 55 miles, and 5 rivers.

From Littlehampton to Dover, 95 miles, there are 8 rivers—

Name.	Area.	Name.	Area.
——	35	Cuckmar	75
Adur	160	Old Haven	121
(Brighton)	56	Rother	312
Ouse	205	——	88

From Dover round by the North Foreland and Whitstable and outside the Isle of Sheppey to Sheerness and up to Chatham, and down the Medway again to Sheerness, and up the Thames to Greenwich to meet the basins of the Thames proper and the Lea, and, crossing over and going down the north shore round by the mouth of the river Crouch to Bradwell, the distance is about 185 miles, and there are the following rivers:—

Name.	Area.	Name.	Area.
Stour	373	Thames and Lea	4,613
The Swale	157	Roding	317
Medway	680	Crouch	181
Cray and others	314		

Proceeding along the east coast to Great Yarmouth, and about 20 miles beyond, to the division of the watersheds of the Bure and Waveney, a distance of 100 miles or thereabouts, the following rivers discharge:—

Name.	Area.	Name.	Area.
Blackwater	434	——	32
——	24	Alde	109
Colne	192	Minsmore	34
——	53	Blyth	79
Stour	407	——	53
Orwell	171	Yare & Waveney	880
Deben	153	Bure	348

Again, to and a little beyond the Withern Eau, at Saltfleet, a distance of about 130 miles—

Name.	Area.	Name.	Area.
Glaven	293	Welland	760
Nar	131	Witham	1,079
Wissey	243	Steeping	101
Ouse	2,607	Withern Eau	189
Nene	1,077		

Passing the mouth of the small river Ludd, and entering the Humber, and proceeding along the south shore past Grimsby, New Holland, and Winteringham, we come to the mouth of the Trent, then the Don, the Aire, and the Yorkshire Ouse; and, on the left bank of the Ouse, the Derwent, and lower down, the Foulness and the river Hull; then, passing round Spurn Point, we come to Hornsea, having traversed the tideway for a distance of 135 miles or so, although the coast-line from Saltfleet to Hornsea is not more than 40 miles; yet we are bound to go so far up the Humber to meet those great rivers the Trent, the Don, the Aire, the Wharfe (the area of which is here included in that of the Ouse), and the Ouse itself. The areas of these river-basins are as follow:—

Name.	Area.	Name.	Area.
Ludd	139	Aire	815
——	39	Ouse	1842
——	122	Derwent	794
Ancholme	244	Foulness	133
Trent	4,052	Hull	364
Don	682	——	206

Resuming, there are, from Hornsea to Redcar, a distance of about 80 miles—

Name.	Area.	Name.	Area.
(Scarborough)	157	Esk	147
		——	100

From Redcar to Tynemouth, 44 miles, there are—

Name.	Area.	Name.	Area.
Tees	708	Wear	456
——	77	Tyne	1,130

From Tynemouth to Berwick-on-Tweed, 60 miles—

Name.	Area.	Name.	Area.
———	31	Coquet	240
Blyth	131	Aln	104
Wansbeck	126	———	129
———	37	———	37
———	18	Till	231

The object we have in view in making these tabular statements is to show, at a glance almost, what a large number of small river-basins adjoin the sea-coast of England and Wales; but without further remark at present upon that point, we proceed to the north-west coast, and follow it and the west coast to Land's End, whence we started.

From the river Line, in Cumberland, to the mouth of the Kent, at the head of Morecambe Bay, there are the outfalls of the rivers of the Lake district of Cumberland and Westmoreland, the distance being about 125 miles to the southern watershed of the river Kent.

Name.	Area.	Name.	Area.
———	21	Calder	28
Line	104	Irt	61
Eden	915	Esk	64
Wampool	78	Duddon	46
Waver	70	———	28
Ellen	72	———	56
Derwent	262	Leven	202
———	11	Kent	255
Ehen	72		

These rivers convey into the sea the surplus waters from Ulleswater, Haweswater, Bassenthwaite lake, Derwentwater, Thirlmere, Crummockwater, Buttermere, Loweswater, Ennerdalewater, Wastwater, Conistonwater, and Windermere.

From the watershed of the Kent, round by Blackpool and up the Ribble nearly to Preston, and back by Southport to Formby, near the mouth of the Mersey, is about 50 miles, along which the following rivers discharge:—

Name.	Area.	Name.	Area
Lune	418	———	7
Wyre	208	Douglas	168
Ribble	585	———	55

Going from Formby up the Mersey about 10 miles above Liverpool, and crossing over to the mouth of the river Weaver, back by Birkenhead to the mouth of the Mersey and round by Hoose and West Kirby to the head of the Dee estuary, and back round Air Point to the coast at Prestatyn, the distance is about 85 miles, and it includes the outfalls of the basins of the following rivers, viz.:—

Name.	Area.	Name.	Area.
Alt	126 }	Weaver	711
Mersey	885 }	Dee	813

Thence along the northern shore of Wales, round by Great Orme's Head and through the Menai Strait along the western coast of Carnarvon and round to Aberdaron, is 90 miles, and the following rivers fall in:—

Name.	Area.	Name.	Area.
Clwyd	319	Seiont	143 }
———	39 }	Soch	33 }
Conway	222 }	Erch	55 }
———	78 }	Dwyfach	48 }

The Isle of Anglesey has a coast-line of 100 miles, and 5 rivers—

Name.	Area.	Name.	Area.
Braint	53	———	69
Cefni	41	Alaw	58
———	47		

From Aberdaron, along the shore of Cardigan Bay to Cardigan Head is 100 miles, including—

Name.	Area.	Name.	Area.
Prysor	141	Wyrai	23 }
Artro	45 }	Arth	31 }
———	3 }	Aeron	52 }
Mawddach	151 }	———	48 }
Dysynni	64 }	Teifi	386
Afon Dyfi	217 }		
Lery	34 }		
———	24 }		
Rheidol	70 }		
Ystwyth	75 }		

From Cardigan Head round by St. David Head, and St. Govens Head, to the mouth of the Bristol Channel at Worms Head, is a distance of 160 miles, in which are—

Name.	Area.	Name.	Area.
Nevern	94	Towy	514
(St. Bride's Bay)	65	Gwendraeth	73
Cleddau	212	Llwchwr	156
(Pembroke)	114		
	61		
Taff	183		

From Worms Head along the South Wales shore by Swansea, Cardiff, Tredegar, and Newport, to the watershed dividing the Wye from the Severn basins, is 100 miles or thereabouts; and there are 12 rivers, viz.:—

Name.	Area.	Name.	Area.
—	66	Rumney	94
Tawe	106	Ebbw	94
Neath	118	Usk	540
Afon	87	—	55
Ogmore	114	Wye	1,609
—	67		
Ely	81		
Taff	198		

Taking the Severn basin properly to terminate where it meets that of the Wye on one side of the Channel and that of the Bristol Avon on the other, near Aust Passage, and returning down the Channel on the Somerset and Devon shore to its mouth at Hartland Point, west of Bideford Bay, there are, in 120 miles, 15 river-basins, including that of the Severn, viz.:—

Name.	Area.	Name.	Area.
Severn	4,350	—	24
Avon	891	—	29
Yeo	106	East Lynn	41
Axe	101	—	47
Brue	197	Taw	455
—	80	Torridge	336
Parret	562	—	10
—	82		

On the Devonshire and Cornwall coast, from Hartland Point to Land's End, there are river-basins of the following areas, in a distance of 110 miles, viz.:—108, 8, 149, 154, and 43 square miles; and another, in which St. Ives is situated, of about 10 square miles.

SECTION X.

Conduits.

WATER may be very well conveyed along a hill-side in an earthenware pipe, in moderate quantities. A 24in. pipe running half full, with an inclination of 5ft. in a mile, will carry a million-and-a-half gallons a day, and $2\frac{1}{2}$ millions when running two-thirds full. But to go to a smaller size, a 12in. pipe with the same inclination would carry 300,000 gallons a day, half full, and 450,000 gallons if running two-thirds full. A 27in. pipe will carry, at the two degrees of fulness mentioned, 2,300,000 gallons, and 3,400,000 gallons, respectively. A 30in. pipe, nearly 3 million gallons a day half full, and $4\frac{1}{2}$ millions if running two-thirds full.

The carrying capacity of any circular pipe may be found for any degree of fulness from the following considerations. If full, or half full, the hydraulic mean depth is half the radius. This is evident, because the cross-sectional area of the pipe is equal to that of a triangle, the base of which is equal in length to the circumference of the circle, and the height to the radius, half of which, multiplied into its base, is the area. The hydraulic mean depth is the area divided by the whole circumference when the pipe is running full of water, and is, therefore, half the radius; and it is the same when the pipe is running half full, for in that case it is half the area divided by half the circumference; but at any height, H, above or below the centre of the pipe, it is as follows—D representing the diameter, and R the radius:—The width at the surface of the water (see Fig. 14) is $W = \sqrt{R^2 - H^2} \times 2$.

The width V is $\frac{D-W}{2}$. From H and V find the length of that part of the circumference in contact with the water above the centre of the pipe, and add it to the lower half of the circumference, thus procuring the wetted border of the section. The length of the two arcs above the centre is the same as that of a single arc the chord of which is 2 H. Find the chord of half the arc $= \sqrt{H^2 + V^2} = C$. Then $\frac{8\,C - 2\,H}{3}$ = the length of the two arcs above the centre, to be added to that of the lower half of the pipe, to find the wetted border, which, being multiplied into half the radius, gives an area, to which is to be added $W \times \frac{H}{2}$ for the whole sectional area of the stream, and this divided by the wetted border gives its hydraulic mean depth.

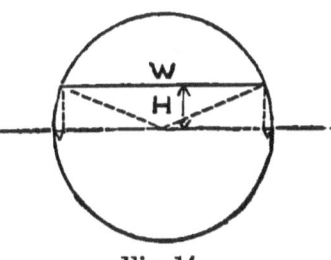

Fig. 14.

For example, in a pipe 24in. diameter running two-thirds full, the height H, is 4in., and the width $W = \sqrt{12^2 - 4^2} \times 2 = 22\cdot 62$.

The width V is $\frac{24 - 22\cdot 62}{2} = \cdot 69$.

The chord of an arc above the centre is $C = \sqrt{4^2 + \cdot 69^2} = 4\cdot 05$.

The length of the two arcs is $\frac{8 \times 4\cdot 05 - 2 \times 4}{3} = 8\cdot 13$.

The lower half of the circumference of the pipe is $\frac{3\cdot 1416 \times 24}{2} = 37\cdot 70$, and $37\cdot 70 + 8\cdot 13 = 45\cdot 83$in. $= 3\cdot 82$ft., the wetted border.

The sectional area is $45\cdot 83 \times 6 + \frac{22\cdot 62 \times 4}{2} =$

320 sq. in. = 2·22 sq. ft., and the hydraulic depth, or mean radius, is $\frac{2 \cdot 22}{3 \cdot 82}$ = ·58ft.

If Eytelwein's rule be adopted, in which h represents the hydraulic mean depth, and f twice the fall per mile, the mean velocity of the stream of water would be $\frac{10}{11} \sqrt{hf} = \frac{10}{11} \sqrt{\cdot 58 \times 10} = 2 \cdot 18$ft. per second, or 130ft. per minute, and the area being 2·22 sq. ft., the quantity is 288 cubic feet per minute, or 2,592,000 gallons per day as before stated.

Larger pipes than those of about 24in. diameter are difficult to handle, require heavy tackle to lift about, and are liable to split longitudinally with external pressure, unless the pipes are evenly bedded all round the lower half, and the haunches of the top solidly filled in between the pipe and the sides of the trench.

At what diameter, exactly, a brick or stone conduit becomes cheaper than a pipe, depends upon the local materials; but, usually, about 20in. becomes the turning-point, in respect of expense, between an earthenware pipe and a culvert of masonry. The pipe has some advantage in being glazed, but for clear water this is not of much importance. A circular conduit, 2ft. diameter and half a brick thick, would usually be laid at less expense than a pipe of the same size. Very excellent conduits are made of square shape, with flag bottom and covers, and sides of clean bedded stones with squared joints. A larger quantity of materials is required for this form than for a circular one, because to carry the same volume of water at the same inclination, a larger sectional area is required, and the walls, being straight, need to be thicker.

The question arises, when considering the advisability of adopting a square or a circular form, what are the comparative carrying capacities of the two forms, the more immediate question being what dimensions of the square form carry the same quantity of water per day or per minute, as a circular conduit of given diameter. This may be seen from the following considerations:—In rectangular channels the best proportion of width to depth of the stream of water is width 2, depth 1. That proportion is the best which gives the greatest hydraulic mean depth with a given quantity of materials used in the construction of the channel, because, the greater that depth, the less need be the sectional area of the stream, the carrying capacity of all conduits having the same inclination being as the area multiplied into the square root of the hydraulic mean depth. A rectangular conduit may be compared with a circular one in the following manner:—Let them both run half full; then if the height of the rectangular conduit be the same as its width, the width of the stream will be twice its depth.

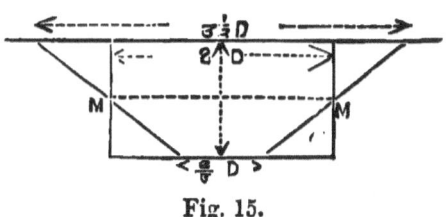

Fig. 15.

Let h represent the hydraulic mean depth, as before, W the width, and D the depth of the stream (see Fig. 15). If the width be twice the depth, the area is $2\,D^2$, and $h = \dfrac{2\,D^2}{W + 2\,D} = \dfrac{2\,D^2}{4\,D} = \cdot 5\,D$; that is, in a rectangular stream in which the width is twice the depth, the hydraulic mean depth is half the simple depth, as, in the circular form, it is half the radius. Then $\sqrt{h} = \sqrt{\cdot 5\,D} = \cdot 707 \sqrt{D}$.

The sectional area multiplied into the square root of the hydraulic mean depth thus becomes $2\,D^2 \times \cdot 707 \sqrt{D} = 1\cdot 41 \sqrt{D^5}$, and this quantity must be the same in the square as in the circular form, to satisfy the conditions.

For example, in the case of a rectangular stream, in which the width is twice the depth, what would be the dimensions to make the carrying capacity equal to that of a circular conduit 3ft. diameter running half full? In this latter form the hydraulic mean depth or mean radius is ·75ft. The area of the lower half of the conduit is $\frac{3 \cdot 1416 \times 3 \times \cdot 75}{2} = 3 \cdot 53$ sq. ft., and the area multiplied into the square root of the hydraulic mean depth is $A \times \sqrt{h} = 3 \cdot 53 \times \cdot 866 = 3 \cdot 05$. Then in the rectangular channel $A \times \sqrt{h} = 1 \cdot 41 \sqrt{D^5} = 3 \cdot 05$. $D^5 = \left(\frac{3 \cdot 05}{1 \cdot 41}\right)^2 = 4 \cdot 66$, and $D = \sqrt[5]{4 \cdot 66} = 1 \cdot 36$ft., the depth, and the width is 2·72ft., the area being 3·70 sq. ft. The border is $W + 2D = 5 \cdot 44$ft., and the hydraulic mean depth is $\frac{3 \cdot 70}{5 \cdot 44} = \cdot 68$ft., being half the depth, as before stated. If the quantities of water carried by the two conduits be calculated from these data, they should be the same.

Small conduits are more liable to freeze than large ones, and should be covered, to protect them from freezing as well as from dirt. Snow is troublesome in an open conduit whether large or small; it clogs the run of water; but the freezing over of the surface of a large body of water is not of so much importance as in a small one. In either case it reduces the sectional area of the stream, and at the same time increases the solid surface with which the water runs in contact, but the effect of this in reducing the quantity of water carried is but little in a large conduit, while it is very considerable in a small one.

In respect of protection from dirt, fencing may be sufficient in the open country, but the expense of fencing bears a greater proportion to the whole expense in a small than in a large conduit, and is almost as much as that of covering a small conduit. If the line of conduit is not subject to contamination by dirt it may be open, though it be not large, if the inclination is sufficient to cause a velocity of about 3ft. per second. This would wash away earth if not protected by a facing.

In an open conduit cut with side slopes at which the ground will stand without slipping, so that the facing is for the purpose of protecting the earth from the wash of water rather than for keeping it up by its weight or lateral resistance, the side slopes would usually be flatter than those which coincide with the best form in respect of carrying capacity. They would usually be 1½ to 1 or more, but side slopes of 1⅓ to 1 are best in respect of carrying capacity, which, with these slopes, is the same as that of a rectangular channel of a width equal to twice its depth, and of the same sectional area. The peculiarity of this slope is that the length of the two slopes and bottom is the same as the length of the two sides and bottom of a rectangular channel of the proportions mentioned, the area of the two forms being the same; consequently the hydraulic mean depth is the same, and, therefore, the discharging capacity. If the slope be drawn through the point M in Fig. 15, it will leave the area of the section the same as that of the rectangular channel, whatever the slope be, but there is only one slope which will leave the length of border the same as that of the rectangular channel; that slope is 1⅓ to 1, and the length of the two slopes is equal to the top width, being 3⅓ times the depth, the bottom width being ⅔ of the depth. Referring to the figure, the area is $\frac{3\frac{1}{3} D + \frac{2}{3} D}{2} \times D = 2 D^2$, and the border is $1\frac{2}{3} D + \frac{2}{3} D + 1\frac{2}{3} D = 4 D$. Hence, the hydraulic mean depth is $\frac{2 D^2}{4 D} = \cdot 5 D$, the same as that of the rectangular channel, and also of the circular one.

But sometimes practical considerations have greater weight than the best deductions, and the sides of an open channel would generally be less troublesome to keep up with slopes of 1½ to 1 or more. Frost is the great disturber of the upper part of open channels, whether large or small, although the covering of water prevents it from reaching the bottom. The sides above the water should be porous—the facing, that is to say, should not be close-jointed—so that water may not accumulate behind it, for when it does so, it swells when frozen and disturbs the facing. But when the facing is thus open-jointed it would be the means of introducing mud into the conduit from behind it in long-continued rains upon clayey ground; to prevent this, the facing should be laid upon a bed of sand.

Through loose ground the making of the conduit water-tight must always be one of the chief difficulties—a difficulty of course to be overcome. For this purpose puddle, to be used with all materials except concrete, is perhaps the least expensive material, and gravel puddle is better for the purpose than clay puddle. A foot is an abundant thickness for small conduits for the bottom and slopes of the ground, and 2ft. above-ground in the embanked portions of the line of conduit. In these the hardest of the excavated earth will be placed along the central portion, and brought up to the level of the underside of the puddling, as far as the height above-ground is but a few feet.

A material which may be used for conduits is concrete, made with Portland cement of good quality. It forms at once a water-tight channel and one which needs but little protection of the surface, a facing of cement being sufficient for all that part under water, but above the water, in an open conduit, it would be preferable to face the concrete with brick or stone, and indeed for some little depth below the water surface. A covered conduit may equally well be made with concrete, both bottom and top, and the thickness

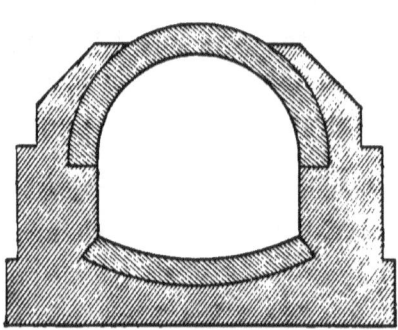

COVERED CONDUIT
Fig. 16.

need not be more than about half as much more as would be the proper thickness of brickwork.

Beyond the several points in a long line of conduit at which slight cutting should end, or the height above-ground would be but a few feet, and where it would be economical to cross straight over low ground rather than extend the line far up the hollow and back again on the other side, the aqueduct is best carried on piers and arches; but economy points to girders, which can be made to carry the water between them on a floor laid upon the lower flanges, or in the form of a wide box-girder; but if the water runs in contact with the iron the conducting property of the metal quickly draws heat from the water in winter. If it be lined with a non-conducting material, which is at the same time brittle, the difference in the rate of expansion by heat of the metal and the lining tends to destroy the adhesion. This may in some measure be prevented, and perhaps entirely, by sinking the aqueduct bodily as far as is necessary to completely immerse it, so that it may preserve a nearly uniform temperature both top and bottom; but this, of course, somewhat adds to the retardation of the flow by increasing the wetted border. To prevent contact with the metal altogether, by substituting an independent carrier for the water within and clear of the girders, would be expensive, but such a carrier might very well be made in earthenware, either rectangular or circular, and of any size by being jointed crosswise.

The least expensive aqueduct would be an earthenware pipe—of stoneware or fireclay—carried on a single girder in saddles upon the top flange. An open girder, continuous over the piers, with wide flanges both top and bottom, would resist the wind, and as the pipe and the water together would be of considerable weight, it would not be likely to be blown off the girder, especially if tied down with a strap over all, every 10ft. or so.

However desirable it may be to follow the contour of

the ground, there will occur, in any conduit of considerable length, breaks in the continuity of such a line, caused by an intervening valley, or by a range of hills, through which a tunnel may be necessary, in order to continue the conduit. In the case of a valley to be crossed, cast-iron pipes are used, laid underground at such a depth that they may have a sufficient covering for their protection (say 2ft.), and rising again on the opposite side to within so much of the level of the conduit from which they depart as will give a sufficient head to the water to force its way through them. The following formulæ by which the necessary difference of level of the two ends of the pipe, or the head, is ascertained, are based on the experiments of Dubuat, Bossut and Couplet. Taking a number of their experiments made with pipes of from 1in. to 5in. in diameter, and from 30 to 1,700ft. in length, M. Prony found that, taking English measures and putting
h = the head of water in feet,
l = the length of the pipe in feet,
d = the diameter of the pipe in feet,
v = the velocity in feet per second = $48 \cdot 49 \sqrt{\dfrac{dh}{l}}$

and by substituting a fall in feet per mile, which call f, $v = \sqrt{\dfrac{df}{2 \cdot 24}}$, and $f = \dfrac{2 \cdot 24 \, v^2}{d}$.

Eytelwein, taking the same set of experiments, found that $v = 50 \sqrt{\dfrac{dh}{l + 50d}}$. This correction $50d$ is for short pipes, but in case of actual waterworks practice it may be neglected, and then Eytelwein's expression may be translated into $v = \sqrt{\dfrac{df}{2 \cdot 11}}$ and $f = \dfrac{2 \cdot 11 \, v^2}{d}$.

In using these and similar rules for practical purposes, engineers have, in exercising a wise precaution in under-estimating rather than over-estimating the capacity of a pipe to deliver water, had a tendency to increase these co-efficients of 2·24 or 2·11, for the purpose of allowing for the obstruction of bends and other irregularities in a long length of pipe, and Mr. Blackwell proposed 2·3 in all cases; but a few of the more eminent water engineers, who have had large experience in gravitation works, have been able to ascertain from actual examples the quantity of water passing through long and large conduit pipes, and Mr. Bateman has said that the very general method of coating pipes inside and out with the pitch of coal-tar, to preserve them from oxidation, has had the effect of diminishing the friction of the water to a considerable degree; so that it is probable that when pipes are so coated the effect of it is to facilitate the passage of the water to quite as great an extent as the bends and other irregularities retard it. Roundly, 10 ft. per mile may be named as the allowance for the difference of level of the two ends of a pipe thus laid across a valley; but where the pipe is large, less than that is sufficient. Ten feet per mile will induce a velocity of 3ft. per second through a 2ft. pipe.

It frequently happens that in a length of several miles of conduit pipe one or more streams of water, or ravines, or other such places, have to be crossed, and instead of passing under them, the pipes are sometimes carried over them, as shown in the illustration (Fig. 17).

Large pipes are now commonly cast in 12ft. lengths, so

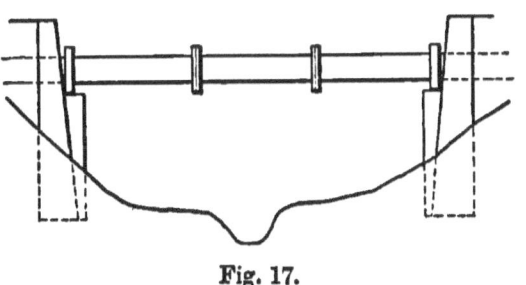

Fig. 17.

that a stretch of 36ft. can be had from side to side, and if a greater length be required, one or more intermediate piers can be built. It is not usual to carry more than three pipes in one stretch; but three are safely thus carried across such places by means of flange joints strongly bolted, the flanges themselves being also of more than the ordinary strength.

As to the strength of cast-iron pipes laid across a valley, it is necessary to consider that concussions may occur if stop valves be placed on the line, and also that concussions may occur if the pipe be so laid as to be liable to air locks, although no concussion would occur from either cause, if in the one the valves were made with a pitch of screw so small as to prevent the valve door being shut down suddenly, and, in the other case, if one or more self-acting air valves were placed on the summit of every rise. The most trying time to a pipe of this kind is when it is first charged with water; but it is easy to admit the water for the first time gradually, and not recklessly to open the valve at the head of the pipe, as if it were already full of water. When these precautions are taken, there is no need to make the strength of a pipe equal to more than six times the strain it is subject to from the head of water upon it; and if we take the low estimate of six tons per square inch as the utmost tensile strength of the iron, we shall have a working strain of one ton per square inch of metal.

At the head of a line of conduit pipe there should be a sluice valve, and a provision made for turning the water sideways into some proper watercourse. The mouth of the pipe should be some feet below the head of water—10ft. is not too much—to prevent air being drawn in to the pipe during the ordinary working. When this valve is shut down, the body of water shut into the pipe will proceed on its course at first with the momentum it had acquired, leaving a vacuum between its head and the sluice valve, which will cause it to partially return, and to oscillate backwards and forwards until the forces within the pipe are balanced.

At every low point a washing-out pipe should be placed,

with a valve upon it, so that the conduit pipe may be emptied when necessary. It is well to have these outlets of limited size, so that carelessness in opening them suddenly may not cause damage to property on the brook course below, if it be a mere brook into which the water is turned.

It will not unfrequently happen that in the course of a long conduit pipe a railway will have to be crossed, and here the pipes should be unusually thick—capable, say, of withstanding ten times the strain they will be subject to.

In cutting the trench for the pipes, air locks should be as much as possible guarded against by cutting through many small elevations of ground, so that long reaches of rise and fall may be obtained. The desirability or otherwise of this certainly depends on the perfection of action of the air valves; for if this action be perfect the more ups and downs the better, for the air has then shorter distances to travel before it can escape; but, considering that a large quantity of air is contained in the water itself which is constantly seeking a higher level, it is perhaps better to afford it as few points of accumulation as possible. The best known air valves are those with a ball of gutta percha falling down from its seat as long as it has nothing but air to support it, the air meanwhile escaping, when pressed by water, but rises to its seat by floating on the water when it reaches it, and as long as no fresh supply of air accumulates about it it is held up tightly to the india-rubber seat provided for it. These valves are made by Messrs. Guest and Chrimes, of Rotherham.

In a discussion on a paper on the Melbourne Waterworks, Mr. Hawksley stated that concussions in mains are caused by the sudden escape of air at the air valves, causing two columns of water to meet each other and give a blow to the pipe, and in one case he ordered the air outlet to be no more than ¼in. diameter, so that the air could escape only gradually.

SECTION XI.

TUNNELS.

ON the Bristol Waterworks there are three tunnels—one at Harptree, about 1¼ mile in length; another at North Hill, about half a mile, and another at Winford of one mile, all on the main conduit, and another on a branch conduit. The first-named of these tunnels was driven through an intensely hard and difficult conglomerate rock, without beds or joints. There were nine shafts, and at some of the faces of the tunnel the driving alone cost £6 per lineal yard; although at others it was only £3. The average cost of powder alone was 10s. per lineal yard. The size of the hole blown out was about 6ft. square, and the very rough bottom and sides were afterwards lined with masonry. The contractor who undertook this tunnel was obliged to give it up before it was half driven, and the driving of it was re-let at £3 14s. per lineal yard, being at the rate of 18s. 6d. per cubic yard, and the masonry at about 30s. per lineal yard, being together £5 4s., but there were extras which brought up the cost to £5 10s. per yard.

The driving of the North Hill tunnel, through magnesian limestone, was estimated to cost £3 5s. per lineal yard in rock, being at the rate of 16s. per cubic yard, and at the loose ends £1 15s. per lineal yard, being at the rate of 7s. per cubic yard.

At the Winford tunnel, through blue lias beds and shale, and some marl, the contractor was paid for the driving of the tunnel in rock £2 10s. per lineal yard, being at the rate of 12s. 6d. per cubic yard; and in clay at £1 15s. per lineal yard, being at the rate of 7s. per cubic yard.

At a water tunnel in Dorsetshire, through the Kimmeridge clay, 5ft. high, and 3ft. 6in. wide in the clear of the masonry, which was 12in. thick, the tunnel being about half a mile in length and 300ft. below the summit of the hill, the driving was estimated to cost, at the loose ends, £1 15s. per lineal yard, and in the main portion of the tunnel £2 10s. It is worth noting that this tunnel gave out so much foul air that the men could not work.

The Alwoodley tunnel of the Leeds Waterworks is said to have cost £6 per lineal yard complete.

The Mottram tunnel, on the Manchester Waterworks, 2,770 yards in length, with four shafts, of which the deepest was 150ft., and with bore holes for air, the ground consisting of clay at one end and sandstone and shale at the other, was estimated to cost per lineal yard:—

	£	s.	d.
6 cubic yards excavation, at 5s.	1	10	0
Brickwork, 1½ cubic yards at 25s.	1	17	6
Centres	0	10	0
Proportion of shafts	0	6	0
Pumping, &c.	0	16	6
Total	5	0	0

The tunnels on the Bradford Waterworks were estimated to cost per lineal yard, where lining was required and also timbering:—

	£	s.	d.
4⅛ cubic yards excavation, at 18s.	3	15	0
1 1/17 cubic yard masonry, at 24s.	1	6	0
12½ cubic feet ashlar in springers and key stones, at 1s. 6d.	0	18	9
Total	5	19	9

and where no timbering was required:—

	£	s.	d.
3 1/10 cubic yards excavation, at 18s.	2	15	9
Masonry as above	2	4	9
Total	5	0	6

SECTION XII.

SERVICE RESERVOIRS.

THE greatest difficulty usually met with in the construction of service reservoirs is that of site and elevation. The remark made by somebody that it was curious how rivers mostly ran through towns must have had its origin in a forgetfulness of the circumstances under which towns have grown up. Towns have begun to be formed on the banks of rivers more or less navigable, but it was the navigability of the river that first induced settlement on its banks for the purpose of commerce; and then, up to the time that steam became known to be so powerful an agent, and as roads began to be made, waterfalls were sought for power for manufactures, as well as for manufacturing uses. So that we find the "centres" of all manufacturing towns close to the rivers running through them; but presently people build their houses outside and away from their factories and warehouses, and therefore on higher ground, and it is this tendency to get to higher ground on which to build houses, that renders the choice of a sufficiently elevated site for a service reservoir difficult to find within a moderate distance of the town. The service reservoirs of Manchester are five miles off; those of Liverpool eight miles; and those of the most feasible scheme once proposed for the supply of water to the metropolis were to be ten miles off. However, those of most other towns are nearer. When the service reservoir is made near the town it is recommended that it be covered, to protect the water from the impurities of the atmosphere; and in all cases a service reservoir should be covered if it is less than 15ft. deep, for if still water be exposed at a less depth than that, vegetation is promoted, and when that dies animalculæ are formed.

The size of a service reservoir, unlike that of an impounding or storage reservoir, depends upon circumstances within the immediate control of the engineer—that is to say, it depends upon the character of the supply. If the supply be through a conduit from a storage reservoir, and the conduit be a short one, one or two days' supply may be sufficient, because anything happening on such a conduit to prevent the delivery of water can soon be put right; while the longer the conduit, or the more complex in character, the greater should be the contents of the reservoir. The service reservoir of the scheme already mentioned as designed for the supply of the metropolis, the conduit of which was to be 180 miles in length, was intended to hold three weeks' supply. Also where a service reservoir is supplied by pumping engines, it should hold a considerable quantity of water, to allow of stoppages of the machinery or part of it. Generally, perhaps from three to seven days' supply may be said to be the ordinary practice.

Reservoirs not requiring to be covered are usually formed by excavating the ground and embanking the earth round it, adjusting the level of the bottom of the excavation so that the quantities of cutting and embankment are equalised. To render the reservoir watertight, it is usual to lay down a flooring of clay puddle, 18in. to 2ft. in thickness, worked in layers in the manner described for the puddle of the storage reservoir. The slopes of the excavation should be not less than 2 horizontal to 1 vertical, and should be cut down in steps, so as to hold up the puddle placed on them. From the top of the slope of the excavation the puddle is continued on the surface of the ground (the bed-puddle) to the centre of the bank, and up through the centre of the bank the puddle is carried as a wall. In reservoirs of considerable size it is necessary to lay down this bed-puddle, or over a part of its area, before the floor-puddle and that on the slopes can be completed, in order to provide a place of deposit for the excavated material, but a margin of it, of not less than 3ft. wide, should be left uncovered, so that the slope puddle may afterwards be joined up to it; and this margin should be formed thus:—

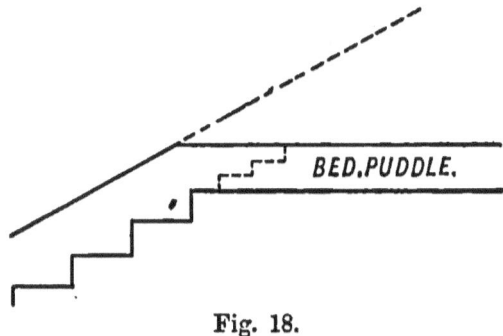

Fig. 18.

So that when the joining up is to be done the layers may overlap and be re-worked into each other. This part of the work requires the greatest watchfulness to ensure soundness.

It is often the case that the thickness of the puddle wall of service reservoirs is made less than that of storage reservoirs proportionally to the height, but it is not very obvious that this is sound practice. It may be said that in the higher bank there are more risks of disturbance; yes, but then the puddle is made thicker in proportion to the height. Of course, if the larger reservoir burst, more damage will be done than if the smaller one were to fail; but this hardly forms a good reason, for the damage done by the bursting of even the smaller reservoir would be sufficiently serious to prevent any risk being incurred on that score. As long as the puddle wall of the larger bank is held up in its place, it has no more force to resist, proportionally to its height, than that of a smaller bank, and the possibility of its not being constantly held up in its place cannot be admitted in the case of the smaller any more than in the larger bank. The only reason of

PUDDLING.

any value why the puddle wall of a large reservoir should be proportionally thicker than that of a smaller one admits a deficiency of practice. It is that in works of magnitude, where large numbers of workmen are employed, it is more difficult to ensure sound work than where the attention of the inspector can be more concentrated; but that only admits the insufficiency of the number of inspectors. And if it be said that there is a difference between the two cases, inasmuch as the service reservoir is generally kept full, while the storage reservoir is subject to fluctuation of level, and therefore of pressure on the puddle wall, this does not seem to touch the question, but rather to be a consideration in the formation of the inner slope of the bank which is to protect the puddle wall.

The bottom should be formed with an inclination from the sides to the centre, and along the centre there should be a channel with an inclination towards the lower end, and from the lower end of this channel there should be a pipe laid to lay dry the bottom of the reservoir—the drain pipe. The level to which the water is drawn off for use is the foot of the slope, some feet above the bottom of the reservoir. The water is drawn off through copper-wire gauze strainers of from 40 to 60 strands to the inch.

For smaller reservoirs, covered over when finished, vertical walls are built, and the excavated material embanked around them. Sometimes these reservoirs are made circular, but they are difficult to cover economically, although the form itself is economical in respect of wall material, the circle containing a greater area than any other form having the same length of wall.

The best way of covering service reservoirs is by running 14in. walls lengthwise at distances of from 15 to 20ft. apart, and turning 9in. brick arches over the spaces, having a rise of one-fifth of the span, filling up the spandrels with concrete, and covering over the whole with earth 18in. or 2ft. in depth. Openings may be made in the walls, either altogether circular, or in the form of vertical spaces arched over, so that the wall consists of a series of piers supporting a continuous depth of wall—say 2ft.—from which the main arches spring. One or more openings in the arches should be made for the escape of air, and a man-hole should be provided, with ladder bars built into the wall, or, which is perhaps preferable, an iron ladder should be provided by which access may be had to the bottom. Ladder bars built into the wall have sometimes been made of cast iron, but at least one serious accident has happened to a workman by the sudden breakage of one of these bars on receiving a blow, and they ought to be of wrought iron.

An overflow weir, capable of passing over the quantity of water that the conduit or the pumping main will deliver into the reservoir, should be placed at the top-water level, and, having fixed upon the greatest height to which the water is to be allowed to rise above this level—say 6in.—the length of the weir may be found (without going into the particulars of the exact form of the weir) by the general equation $l = \dfrac{q}{5 \sqrt{d^3}}$ where $q =$ the quantity of water in cubic feet per minute, $d =$ the depth in inches measured from a still head, and $l =$ the length required, in feet.

The sluice valve, by which the water is discharged from the service reservoir, belongs to the main, which forms the first feature in the means of distribution.

SECTION XIII.

Pressure, and its Effect in Pipes.

The force and mode of action of water under pressure along and at the end of a main pipe may be worth considering in connection with its storage in reservoirs such as have been described. The pressure of still water is the weight of a vertical column of it above the place where the pressure is measured; but still water does no work, and when the column is in motion the pressure is less. The weight of water is as follows, the foundation being the Troy grain, 5,760 of which used to make a pound weight; but this pound not being satisfactory for general purposes in England, the pound weight was increased to 7,000 of those grains. An Act of Parliament made a gallon of distilled water at the temperature 62° F. to weigh 10lb., or 70,000 grains, in air of the density produced by a pressure equal to the weight of a column of mercury 30in. high; and established also, from experiments which had been made by a commission, that a cubic inch of distilled water at the temperature and pressure above mentioned weighs 252·458 grains. A cubic foot therefore weighs $252·458 \times 1{,}728 = 346{,}247$ grains, or $\dfrac{436{,}247}{7{,}000} = 62·321\text{lb}$. In half a dozen different tables of the weight of mercury compared with that of water, six different values may be seen—viz., 13·56, 13·568, 13·57, 13·58, 13·596, and 13·6. When tables differ, what is the proper weight? The difference seems to arise from comparing the weight of mercury at one temperature with that of water at another, in some cases, and in other cases in taking water sometimes at the temperature 39·2° F., when it is at its greatest density, and at other times taking it at the common temperature of 62° F. This latter is the more useful for ordinary purposes. At 62° F. mercury is 13·596 times heavier than water. The pressure of the atmosphere then is the same as a column of water $\dfrac{30 \times 13 \cdot 596}{12} = 33 \cdot 99\text{ft.}$ high, or 34ft., and the corresponding pressure per square inch is $\dfrac{34 \times 12 \times 252 \cdot 458}{7,000} = 14 \cdot 71\text{lb.}$ When the mercury rises to 30·5in. the pressure is 15lb. nearly per square inch. But as the weight of the atmosphere is often less than 30in. of mercury, and sometimes only 28·5in., it is the minimum pressure which should be reckoned upon in practice in order to guard against failure of action at all times. This is $\dfrac{28 \cdot 5 \times 13 \cdot 596}{12} = 32 \cdot 29\text{ft.}$ of water, and the corresponding pressure per square inch is 13·97lb. or 14lb. As the head of the column of water is open to the atmosphere equally with the point at which the pressure is applied, this pressure is "above the atmosphere," and although there may be a little difference between the pressures of the atmosphere at the two ends of the column, measured at the same instant, if the pressure reckoned upon be that of the minimum there will be no failure of effect.

The weights above mentioned relate to water without admixture of other matter; common water contains in solution heavier matter than water itself, derived from the ground through or over which it runs, and in practice a cubic foot is taken to weigh 62½lb. and to contain 6¼ gallons.

In the case of a pipe conveying water from a reservoir to the place where it is to be used, the column of water divides itself into two portions, and the pressure at the lower end of the column is according to the height of the lower portion. The upper portion, being that part of the whole column between the top of the virtual column and the level of the water where it meets the atmosphere, is the head of water, which feeds the column as fast as it descends by the force of gravity. In a pipe of any given diameter, as 1ft., the resistance of the sides of the pipe to the motion of the water is proportionate to its length, for it is proportionate to the area of surface with which the water runs in contact; secondly, it is proportionate to the velocity of the water, for it is proportionate to the number of particles of water in contact with a given length of pipe, as 1ft., during any given time, as one second. But the head of water required to supply the force of which the column of water is robbed in overcoming this resistance must be as the square of the velocity, for not only is it thus simply proportionate to the velocity with which the water moves, but it must be replenished as fast as it runs away. If the length of pipe in contact with the water during one second be twice as great in one case as in another, then there will be twice the resistance due to that cause, requiring twice the head to overcome it, and at the same time the water will run away from the head twice as fast, and require twice the length of pipe full of water to replenish it in the same time.

The head, therefore, is as the square of the velocity. Now as the altitude of the proper head cannot be increased, but remains at the same level, nearly, whatever the velocity in the pipe may be, the only way in which the head can be increased is by taking from the length of the real column the height which may be necessary to give the required velocity—increasing the head at the expense of the real column, and thereby reducing its effective pressure. The same effect may be shown by substituting actual pressures for heads of water in a horizontal pipe, such as a pumping main. At a point on a line of main let three sections of it be marked off, of equal length, say 1ft., in the direction in which the water flows, and let these sections be numbered 1, 2, and 3 from the point of observation. From the same point let three other sections be marked off, of the same length, in the direction from which the water comes, and let these be called A, B, and C, and let them be bodies of water. Let the diameter of the pipe be such that each body of water requires a pressure of 1lb. to be given to it to overcome the resistance of each of the sections 1, 2, and 3. Then let three operations be performed:

1. A moves through section 1, requiring a pressure of 1lb. 1
2. A and B move through sections 1 and 2, A requiring 2lb. and B 2lb. 4
3. A B and C move through sections 1, 2, and 3, A requiring 3lb., B 3lb., and C 3lb. 9

Then if each operation is performed in the same time, as one second, the pressure required is as the square of the velocity.

If feet vertical of water be substituted for pounds pressure, the illustration serves equally for a head of water.

In respect of diameter: in pipes of different diameters the resistance is inversely as the diameter. The resistance is, indeed, directly as the area with which the water runs in contact, which in the same length of pipe is as the circumference, and therefore as the diameter, and a diameter of two offers twice as much resistance as a diameter of one; but the distance at which it acts—the radius of the pipe—is twice as far removed from the axis (the point in which there is no resistance), and this, be it remarked, in two opposite directions, so that the effect of distance is as the square of the radius, and therefore of the diameter; the resistance being direct in its action, and the distance at which it acts inverse, the result is that the resistance is inversely as the diameter.

These several forces and resistances are formulated thus:—

Let v = velocity per second.
l = length of pipe.
d = diameter.
h = head of water.

Then $v^2 = \dfrac{c\,d\,h}{l}$

in which c is a constant multiplier deduced from experiments to be 2,500 when all dimensions are taken in feet. This is the value of c according to Eytelwein's deduction for long pipes. The velocity, then, in feet per second is $v = \sqrt{\dfrac{2{,}500\,d\,h}{l}} = 50\sqrt{\dfrac{d\,h}{l}}$; and the diameter is $d = \dfrac{v^2\,l}{2{,}500\,h}$. This is the usual requirement in the case of a pipe to convey water from a reservoir to the place where it is to be used, for in such case the height and length are fixed, and the velocity must be limited so that at its maximum it may do no harm to the pipe by its violence, and so that branches derived from it may be duly filled, and so that a sufficient working pressure be given.

When the pipe is several miles in length the rule is converted into a form expressing the head of water in feet per mile by dividing 5,280 by 2,500, in which l is eliminated, and $h = \dfrac{2 \cdot 11\,v^2}{d}$. If the velocity be made 3ft. per second, $h = \dfrac{19}{d}$, and the pressure of water running with that velocity would be, at the distance of one mile from the reservoir, less than that due to the whole height of the column by $\dfrac{19}{d}$.

It might seem that the pressure would be that due to the whole height less such a head as is required to produce the velocity considered as falling water merely, which would be $\dfrac{v^2}{64}$; but although that is so in open streams, which are in train, it is not so in pipes under pressure. In this case, as in the other, there is a certain gradient line, which, if drawn straight between the head of water in the reservoir and the height due to the pressure at the end of the pipe, is the gradient which that pipe would necessarily take if it ran full, but only full, that is without pressure on the highest part of its circumference; and the pressure at any part of the length of a pipe is that due to the vertical height between the actual position of the pipe and that gradient, which is the "hydraulic gradient." There being thus a gradient line in all conduit pipes, there must be a length at which there would be no pressure, supposing the pipe to be prolonged so far from the reservoir, and there the water would simply run out of the end of the pipe with the velocity due to the gradient. This length is found as follows:

Substituting H, the whole head, for h, the head which is due to the velocity v,

$$l = \frac{2{,}500\; d\; H}{v^2}.$$

For example, let $d = 2$ft., $H = 100$, $v = 3$, then $l = \dfrac{2{,}500 \times 2 \times 100}{9} = 55{,}555$ft. Thus at the distance of 10 miles or so there would be no pressure in a pipe 2ft. diameter and 100ft. below the reservoir, and it could not carry the water farther than 55,555ft. with a velocity of 3ft. per second; beyond that distance the velocity would diminish, unless an actual gradient were immediately given to the pipe equal to the hydraulic gradient; but that being done, the same velocity would continue to any distance, if the diameter remain the same; the stream of water through the pipe would then be "in train," the forces acting upon it being balanced by the resistances, and it would flow under the same conditions as a river. But if pressure were required, there would need to be either a greater head, a shorter length, or a larger pipe. The conditions to be determined beforehand are (1) the quantity of water required, (2) the pressure required, (3) the height at which it is to be supplied above a fixed datum level; and (4) the height of the reservoir above the same datum.

In one of the examples previously referred to, the quantity of water for supply was found to be 3,000,000 gallons a day, after leaving to the stream one-third of the total available quantity. If the distance at which this quantity were required to be delivered were 5 miles, and the height of the storage reservoir above the service reservoir in which the water would be delivered were 150ft., the following conditions would ensue. The quantity per second corresponding to 3,000,000 gallons a day is $\frac{3,000,000}{540,000} = 5\cdot 55$ cubic feet. If the velocity be limited to $2\frac{1}{2}$ft. per second, the diameter would be $d = \sqrt{\frac{5\cdot 55}{2\cdot 5 \times \cdot 7854}} = 1\cdot 68$ft., say 21in. The head would be $h = \frac{5,280 \times 5 \times 6\cdot 25}{2,500 \times 1\cdot 68} = 39\cdot 29$ft., say 40ft. The effective head would be $150 - 40 = 110$ ft., and the pressure per square inch $\frac{110 \times 62\frac{1}{2}}{144} = 47\cdot 7$lb. at the end of a 21in. main 5 miles long.

Upon a line of conduit pipe such as this there may be taken off one or two branches. Say that one-sixth of the water is wanted at three miles distance from the reservoir, and another sixth at four miles, leaving 2,000,000 gallons per day to go to the far end. If the same velocity were preserved throughout, the square of the diameter of the pipe would be reduced one-sixth at the first branch, and one-third at the second. Thus the first length of three miles being 21in. diameter, the second length of one mile would be $19\cdot 18$, or, say, 20in.; and the third length of one mile $17\cdot 14$, or, say, 18in. Each branch takes off 500,000 gallons a day, and, preserving the same velocity, the diameter of each would be $\sqrt{\frac{(21)^2}{6}} = 8\cdot 51$in., or, say, 9in. The effective head in the main conduit pipe at the point where the first branch is derived would be found by deducting the loss of head due to its distance from the reservoir, three miles, which would be $\frac{3 \times 40}{5} = 24$ft., from the difference of altitudes of the reservoir and the branch above the datum. For the second branch the deduction would be rather greater per mile, the diameter being less, while the velocity is the same, for the head is inversely proportionate to the diameter, according to the formula $c\ h\ d = v^2\ l$. Thus if the loss of head due to a velocity of $2\tfrac{1}{2}$ft. per second in the 21in. pipe be 8ft. per mile, it would be in the 20in. pipe $\dfrac{21 \times 8}{20} = 8 \cdot 4$ft. per mile, and in the 18in. pipe $\dfrac{21 \times 8}{18} = 9 \cdot 4$ft. per mile. Adding together the losses of these three lengths—viz., for the first three miles, 24ft., for the fourth mile, $8\tfrac{1}{2}$ft., and for the fifth mile, $9\tfrac{1}{2}$ft. the total loss at the end of the five miles would be 42ft., instead of 40ft., as it would be if the full diameter were continued to the end, and the quantity delivered were 3,000,000 gallons a day.

Cast iron is the material for water-pipes, from 3in. to 3ft. 6in. diameter. They have been cast as large as 44in. and 48in., but there are several reasons of convenience and economy which make it desirable to limit the diameter of cast pipes to about 40in. or 42in. If the quantity of water to be conveyed is such as to require a larger diameter, it would probably be found that either two cast pipes, or one pipe built up of rolled plates, would be preferable. They are sometimes made smaller than 3in.; but the expense of laying constitutes a considerable part of the whole expense of small pipes, and it is not much more for 3in. than for 2in. pipes.

The metal of which pipes are cast is not so strong as that put into girders and some other structures. The tensile strength of this metal is about eight tons per square inch; but that put into pipes is probably not more than seven tons per square inch for many of the pipes of every set, and it is the strength of these upon which the success of the work depends. Whatever the quality and ultimate strength of the metal may be, the working strain should not exceed a determinate part of it, arrived at in two stages of the process of calculation.

In the first place, every pipe is subjected to an actual

water pressure before it is laid. If the pressure be applied steadily to represent the effect of a head of water, it may be as much as one-half of that which the metal would bear before breaking; but if, while under pressure in the proving press, the pipes be struck with a hammer to represent the jarring and blows to which they are subject in the ground, the pressure applied should not be more than one-third of the ultimate strength of the metal.

The proof strain of the pipes in the press should not, on the one hand, be so great as to injure the strength of the metal, which would probably happen if more than half the pressure due to the ultimate strength were applied; but, on the other hand, the actual strain produced in the metal should be sufficient to prove the soundness of the casting, and the quality of the workmanship.

But between the time of proving and being laid in the ground the pipes are subject to a variety of accidents which tend to produce defects, some of which may be undiscoverable; and, moreover, the arrest of the motion of the water passing through them produces a greater strain than that due merely to the height of the column of water. A complete stoppage of the motion could not be suddenly effected, but there is no doubt that the shutting of stop-cocks, in a system of piping, throws additional pressure on some parts of it; and the working pressure, due to the head of water, should not be more than some determinate part of that applied in proving the pipes—say one-third of the pressure applied in the first-named manner, or one-half applied in the manner secondly named; so that, in calculating the thickness of metal, the steady pressure due to the head of water should not exceed one-sixth of the ultimate strength of the metal. If this be seven tons, or, perhaps, not more than 15,000lb. per square inch, the strain produced by the head of water should not exceed $\dfrac{15,000}{6} = 2,500$lb. per inch.

If $H = $ the height of the column of water in feet, the

pressure per square inch on the internal surface of the pipe is $\dfrac{62\frac{1}{2} \text{ H}}{144} = \cdot 434 \text{ H}$.

This is a radial pressure acting on the whole circumference of the pipe equally all round, except that it is less on the top of the pipe than the bottom, by the pressure due to the diameter; but this may be neglected, and its consideration may, in fact, be done away with, by measuring the height of the column of water from the bottom of the pipe. But, although the pressure is radial, and the circumference receives a nearly equal pressure per square inch all round, the component part of the pressure which constitutes the force tending to tear the metal asunder, acts in a direction tangential to the circumference at any point, and at right angles to the radius; and as all the radii are alike in a circular pipe, the force tending to burst the pipe is proportionate to the radius, and is measured by it multiplied into the pressure due to the head of water. For calculation of the thickness of metal, it answers all purposes to take one inch of the length of the pipe.

The whole outward pressure in pounds, tending to separate one-half of the ring from the opposite half, is ·434 H multiplied into the diameter in inches, and this force is restrained by the two sides of the pipe. As it is only necessary to take into account one side, the strength may be calculated from the radius instead of the diameter. Let ·434 H = p = the pressure of the water in pounds per square inch; r = the internal radius of the pipe in inches; t = the thickness of the metal, and s = the strain per square inch of the metal acted upon, to which it is subject under the pressure p and radius r, then the following quantities should be equal to each other, viz :— $p\,r = st$, and $t = \dfrac{p\,r}{s}$

But here is an anomaly; for the part of the metal which the pressure acts upon with the force $p\,r$ is infinitely thin. Beyond the face, the metal is strained less and less as the radius increases.

STRENGTH OF CAST-IRON PIPES.

The following remarks depend upon the truth of the law that the extension of a material per unit of its length, strained by an elongating force, is proportional to the force applied. The extension of cast iron, under all strains up to that which breaks it completely, is not exactly proportional to the force applied; but it is so within the limits of the smaller strains to which the metal is subject in water-pipes.

The strength with which any portion of the metal restrains the force which tends to tear it asunder, is measured by the extension of the metal per unit of its length within the limits of its proper elasticity, and both the force and that extension diminish as the radius increases. The metal at the back of the pipe can only assist the strength of that at the internal face, in proportion to the rate of extension which the force within the pipe produces in it; and the rate of extension at any part of the thickness of the metal is inversely proportionate to its distance from the centre of the pipe.

The magnitude of the bursting force is itself proportional to the internal radius, and is confined to it, and its proportional stress at any part of the thickness is inversely as the distance at which it acts; and as the rate of extension is also inversely as the same distance, therefore the metal, at any part of its thickness, restrains the bursting pressure with a force which is inversely proportionate to the square of the radius at that part.

With an internal radius of 1, a thickness of 1, and a force of 1, the back of the pipe is strained with a force of $\frac{1}{2}$, and as the length of its circumference is twice that of the internal circumference, the extension due to a force of 1 would there be $\frac{1}{2}$; but as the force is only $\frac{1}{2}$, the strain there is $\frac{1}{2}$ of $\frac{1}{2}$, or $\frac{1}{4}$; that is, it is inversely as the square of the radius.

In a 10in. pipe, 1in. thick, the internal radius is 5in. and the external radius 6in., and with any given pressure of water the back of the pipe is strained less than the face in the ratio $5^2 : 6^2 = 25 : 36$.

For the sake of showing the effects in an extreme case, we may take the metal to be still thicker. The distance at which the force acts directly on the metal is limited to a radius of 5in. As the force does not act directly upon the metal beyond the face, but indirectly through the interposing metal, it is diminished as the radius increases. At 6in. radius the force is $\frac{5}{6}$ths of whatever it is at 5in.; at 8in. radius it is $\frac{5}{8}$ths, and so on; and the effect of this diminishing force is further reduced in the same ratio by the distance at which it acts, so that at 6in. from the centre of the pipe there is a force of $\frac{5}{6}$ths acting at a distance of 1 $\frac{1}{5}$th; at 7in. a force of $\frac{5}{7}$ths acting at a distance of 1 $\frac{2}{5}$ths, and so on.

Whatever strain per square inch it may be thought proper to subject the metal to at the interior of the pipe, that at the back will be in the following proportion. Let the maximum strain be assumed to be 2,500lb. per square inch, then,

At 5in. radius (face of the metal) $\dfrac{2{,}500 \times 5}{5} = 2{,}500\text{lb.}$

,, 6in. ,, ,, ,, ,, $\dfrac{2{,}500 \times 5}{6 \times 1\frac{1}{5}} = 1{,}736\text{lb.}$

,, 7in. ,, ,, ,, ,, $\dfrac{2{,}500 \times 5}{7 \times 1\frac{2}{5}} = 1{,}276\text{lb.}$

,, 7½in. ,, (middle of the thickness) $\dfrac{2{,}500 \times 5}{7\frac{1}{2} \times 1\frac{1}{2}} = 1{,}064\text{lb.}$

,, 8in. ,, ,, ,, ,, $\dfrac{2{,}500 \times 5}{8 \times 1\frac{3}{5}} = 976\text{lb.}$

,, 9in. ,, ,, ,, ,, $\dfrac{2{,}500 \times 5}{9 \times 1\frac{4}{5}} = 772\text{lb.}$

,, 10in. ,, ,, ,, ,, $\dfrac{2{,}500 \times 5}{10 \times 2} = 625\text{lb.}$

The mean strain (M) of the whole thickness is not $\dfrac{2{,}500 + 625}{2}$, but is the strain (S) at the face of the metal multiplied into the square of the internal radius, and divided by the mean of the squares of the internal radius (r) and the external radius (R).

Thus $M = \dfrac{2r^2 S}{R^2 + r^2}$

In the case above stated it is $\dfrac{2 \times 25 \times 2{,}500}{100 + 25} = 1{,}000$ lb. per square inch.

The force producing these strains is pr = the pressure per square inch, multiplied into the internal radius of the pipe in inches. In order that the inner portion of the metal be not overstrained, it is necessary to extend the thickness far enough to comprehend a restraining force equal to that exerted on the inner portion of the metal, and its limit is found at a radius the square of which bears the same relation to the square of the internal radius as the force exerted at the face of the metal bears to that exerted at the back, and in very thick pipes the mean strain per square inch of the whole thickness differs widely from the arithmetical mean between the strains at the internal and external radii, because the strain at every part of the thickness is as the square, inversely, of its distance from the centre of the pipe, but at all distances which lie within the thickness of ordinary cast-iron pipes the mean strain would not differ much from the arithmetical mean between the inner and outer strains.

In the diagram Fig. 19, A B represents the strain per square inch at the face of the metal, and C D, E F, &c., the strains at the several distances C, E, &c., from the centre of the pipe, the corresponding strains being inversely as the squares of those distances, in pounds per square inch, produced by the force $p\,r$.

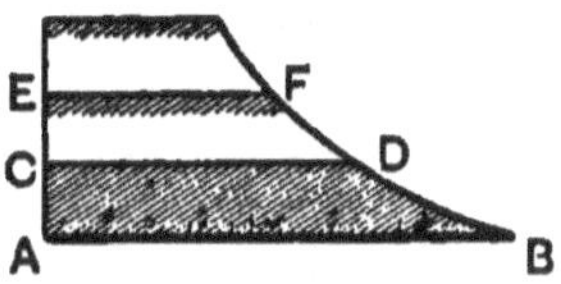

Fig. 19.

There are many short rules for the thickness of metal in cast-iron water-pipes, and they vary much in the results obtained from them. They may be compared with the principles above stated. Mr. Molesworth, in the well-known 'Pocket-book of Engineering Formulæ,' gives the rule,

$t = {\cdot}000125\,p\,d + {\cdot}37$ for pipes less than 12in. diameter.

or,

$t =$ the same $+ \cdot 50$ for pipes from 12in. to 30in. diameter,

t being the thickness of metal, p the pressure of water per square inch, and d the diameter in inches. To adopt a uniform notation, let the radius (r) be taken instead of the diameter, then

$t = \cdot 00025\, p\, r + \cdot 37$ for pipes less than 12in. diameter,

or,

$t =$ the same $+ \cdot 50$ for pipes from 12in. to 30in. diameter.

Mr. Neville, in 'Hydraulic Tables, Co-efficients, and Formulæ,' gives the rule for pipes cast vertically,

$$t = \cdot 0016\, n\, d + \cdot 32,$$

and for pipes cast horizontally,

$$t = \cdot 0024\, n\, d + \cdot 33,$$

in which n is the number of atmospheres of pressure, $=$ the head of water in feet divided by 33. Adopting, likewise, r instead of d, and converting n into pounds per square in. (p), the rule would be, for pipes cast vertically,

$$t = \cdot 0002234\, p\, r + \cdot 32,$$

and for pipes cast horizontally,

$$t = \cdot 000335\, p\, r + \cdot 33.$$

But Mr. Neville remarks that, in practically applying these formulæ, the value of n should always have ten added to it. He mentions a formula adopted by M. Dupuis, the engineer of the Paris Waterworks, which is $t = (\cdot 0016\, n + \cdot 013)\, d + \cdot 32$.

Taking in this case also the radius instead of the diameter, and converting n into pound per square in. the rule would be $t = (\cdot 0002234\, p + \cdot 026)\, r + \cdot 32$.

Professor Rankine, in his 'Manual of Civil Engineering,' gives the rule $\dfrac{t}{d} = \dfrac{H}{12{,}000}$, in which H is the head of water in feet. This would be equivalent to

$$t = \cdot 000192\, p\, d$$

or, $\qquad\qquad t = \cdot 000384\, p\, r.$

But Dr. Rankine remarks that shocks from without

cause the thickness of cast-iron pipes to be often made considerably greater than that given by the above rule. The following empirical rule, he says, expresses very accurately the limit to the thinness of cast-iron pipes in ordinary practice—viz., that the thickness is never to be less than a mean proportional between the internal diameter and $\frac{1}{48}$th in.

Mr. Thomas Box, in 'Practical Hydraulics,' gives as an empirical rule,

$$t = \left(\frac{\sqrt{d}}{10} + \cdot 15\right) + \left(\frac{Hd}{25,000}\right)$$

t, d, and H being the same expressions as those before stated, and in the same terms.

It is to the investigation of Professor Peter Barlow that we are indebted for a solution of the difficulty of apportioning to cast-iron pipes a thickness properly proportionate to the strain upon them from internal pressure. Barlow's rule is $t = \dfrac{p\ r}{c-p}$, in which t, p, and r are the same expressions as those used above, and $c =$ the cohesive strength of the metal per square inch, which is assumed above to be 15,000lb. before it breaks, but only 2,500 as a maximum working strain on the inner portion of the metal, and it is only metal of good quality which it would be safe to subject to even this degree of strain—in metal, that is to say, produced from "mine" or native ironstone; with any considerable admixture of inferior metal the pipes would probably not bear straining nearly so much. Professor Barlow's investigation did not extend to the minutiæ of foundry necessities, such as increase of thickness for unavoidable defects in casting, or the difficulty of making very thin castings; but it established a true basis around which these practical requirements gather, and to which they are more or less wisely added.

To compare these various rules, let each of four pipes—viz., 10in., 14in., 20in., and 30in. diameter—be subject to the same pressure, viz., 174lb. per square inch (400ft. head).

Diameter.	10in.	14in.	20in.	30in.
$p\,r$	870lb.	1,218lb.	1,740lb.	2,610lb.
	t	t	t	t
Barlow	·374	·520	·748	1·12
Rankine	·334	·468	·668	1·00
Molesworth	·588	·804	·935	1·15
Neville	·674	·815	1·02	1·38
Dupuis	·644	·774	·969	1·29
Box	·569	·673	·805	·986

According to Professor Rankine's rule to limit the thinness of pipes to a mean proportional between the internal diameter and $\frac{1}{48}$ in., that would make the least thickness $t = \sqrt{·0208 d}$, and for

  10in. pipes would be ·456
  14in. ,,   ,, ·539
  20in. ,,   ,, ·645
  30in. ,,   ,, ·790

The thicknesses above stated by Mr. Neville's rule are those of pipes cast vertically. The increased thickness required by his rule for pipes cast horizontally is due to remediable defects, and we must allow to all the other formulæ that they demand strictly proper workmanship, and make no allowance for defects which are not absolutely unavoidable; moreover, casting pipes vertically, of nearly all sizes, is becoming more and more the custom. It may be remarked upon Mr. Neville's rule that it is not necessary to add 10 to the value of n (although this is done in the above table) unless it be for very small pressures.

For facility of application, and near agreement with actual practice, Box's rule is preferable to any of the others given above.

SECTION XIV.

AQUEDUCTS.

OF the form and construction of *large* conduits, or aqueducts, there are no better examples than those proposed for bringing water to London from the Lake district by Mr. Hemans * and Mr. Hassard,† and from Wales by Mr. Bateman. Neglecting altogether the merits of either of these projects as a water supply for London, in respect of the source whence it was to be derived, the form and construction of the aqueducts may be usefully examined in connection with the present subject, especially as these were nearly alike in both the cases named, and were such as almost any large conduits or aqueducts must probably have been in this country; nor need the same form and construction be confined to particularly large conduits, for, in any one of considerable length, whether of large size or not, the same features would be met with. Unlike a railway, which may go up a hill or down a valley, and also a canal, which, with locks, may be made in a similar manner, a conduit for conveying water by gravitation must have its regular inclination, not necessarily the same in all parts of its length, but the falling gradient must be continuous.

In each of these aqueducts there were six chief features, viz.: (1) tunnels, (2) open watercourses, (3) "cut and cover," that is, where the ground is cut open and the conduit constructed and arched over, and the earth filled in over the top; (4) raised aqueducts over ground lying

* The late George Willoughby Hemans, M.Inst.C.E.
† Richard Hassard, Esq., M.Inst.C.E.

below the general gradient of the conduit; (5) the portions connecting the two forms of construction, 2 and 4, or 3 and 4; (6) pipes laid underground across valleys too wide and deep to be crossed at the gradient height, rising again on the other side to the proper level, and the conduit being continued in one or other of the ways indicated by 1, 2, or 3.

It was not the first time the engineers above-named had had to consider how best to convey a large quantity of water a long distance. Mr. Bateman had then recently made the works which supply Glasgow with water from Loch Katrine, and Mr. Hassard had laid down the plan for supplying Dublin from the river Vartry. The aqueduct to convey water to London from Ulleswater was to carry 250,000,000 gallons per day. The gradient was to have been 6in. per mile in some parts, and less than that in others, the principle laid down on the score of economy being that the greater fall, and therefore the smaller sectional area, should be given to the tunnels and aqueducts, which are the most expensive portions, and the lesser fall and larger dimensions to those parts which are of comparatively cheap construction, viz., the open watercourses and those where the ground is cut open and filled in again, and where useful material would be got from the excavations.

In most conduits of considerable length it will be advantageous to save head, so as to deliver the water at the far end at as high an elevation as possible, and the value of this may often be greater than additional expenditure required for a larger conduit. The value of each can be strictly calculated.

With 4in. fall per mile in the open aqueduct shown in Fig. 20 (see next page), which is reduced from the sections given in the Appendix F of the Report of the Metropolitan Water Supply Commission, 1869, the mean velocity was calculated at 109ft. per minute, and the quantity of water discharged at 28,340 cubic feet per minute, the width of the stream being 30ft. at the surface

LARGE CONDUITS. 115

and 20ft. at the bottom. Depth in the middle of the stream, 10ft. 6in. Side slopes ½ to 1. Bottom, at the centre, 6in. lower than at the sides. With a fall of 6in. per mile the surface width would be reduced to 26ft., and

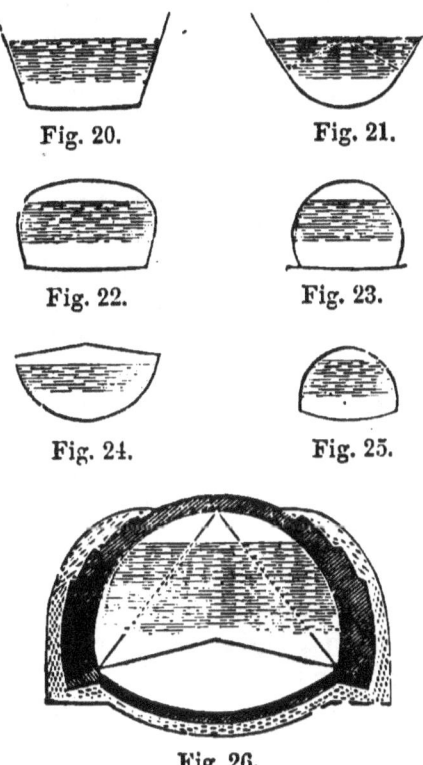

Fig. 20. Fig. 21.

Fig. 22. Fig. 23.

Fig. 24. Fig. 25.

Fig. 26.

the bottom width to 16ft., the depth being 10ft. 6in. at the middle. With these dimensions and fall the mean velocity was calculated at 128ft. per minute, and the discharge at 27,702 cubic feet per minute. These are the sections and form of construction in rock, which is for the most part the Silurian slate formation.

In ordinary ground the form was to have been as shown in Fig. 21. With this form, and a fall of ·4ft. per mile, the velocity, it was said, would be 119ft. per minute, and the discharge 28,760 cubic feet, the width at the surface of the water being 28ft. and the depth 12ft. The curved bottom extends to about half the height, at which point the straight sides touch the circle tangentially. With a fall of 6in. per mile the dimensions would be reduced to a width of 26ft. at the surface, and a depth of 11ft. 6in. With these dimensions and fall the velocity would be 131ft. per minute, and the discharge 28,361 cubic feet.

In the covered aqueduct shown in Fig. 22, with a fall of 4in. per mile, the velocity would be 106ft. per minute and the discharge 28,514 cubic feet. The width at the bottom is 22ft. and at the springing of the arch, 7ft. above the bottom, it is 24ft. The depth of water is 12ft., or 5ft. above the springing. Where the fall would be increased to 6in. per mile, the bottom and middle widths would be reduced to 18ft. and 20ft. respectively, the depth remaining 12ft. With these dimensions and fall the velocity was calculated at 127ft. per minute, and the discharge at 28,575 cubic feet per minute.

In the tunnels shown in Fig. 23, with a fall of 6in. per mile, the velocity would be 125ft. per minute, and the discharge 28,240 cubic feet, the width at the bottom being 18ft., the middle width 20ft. at a height of 5ft. above the bottom, the arch semicircular, and the surface of the water 7ft. 6in. above the springing, the depth of water being 12ft. 6in. No puddle was proposed to be used in any part of this aqueduct, but the excavations were to be lined with concrete, about 18in. thick, faced with rubble stone where the ground is rock, and in ordinary ground the concrete would in some parts be faced with brick on edge, and in others rendered with Portland cement.

With respect to the water-tightness of the aqueduct, Mr. Hassard said that in the construction of a great portion of the open aqueduct, it was probable that no lining would be required. This evidence is similar in effect to that of Mr. Bateman, who, having then recently made the aqueduct bringing water from Loch Katrine to Glasgow, said that the ground there consists chiefly of the mica-slate and clay-slate formations, and that very little lining was required; it is almost impervious to water. There was no clay along that line, and no means of transporting materials, and, knowing the geological formations of the country, Mr. Bateman had adopted tunnels instead of looking for a line along the surface, which had been done by others without success when turning their attention to that source for the supply of the city.

There are on this line of aqueduct as many as seventy different tunnels, of comparatively short length; no long tunnel, except the first one, of 2325 yards, on leaving Loch Katrine. Tunnelling, he said, where short tunnels can be made, and where the materials can be brought to daylight at short intervals, is perhaps the cheapest and the best and the quickest way of getting through a country of that sort. After passing through the first tunnel, the water is conveyed for six or seven miles along the hillsides, not, in this case, in an open aqueduct, but in the manner called "cut and cover." Where the conduit passes through an agricultural country it may be open; there is rather an advantage than otherwise in its being exposed; but near towns or through a thickly-wooded country, or on slopes of mountains liable to snow-drifts, the conduit should be covered. In any case the drainage of the hillsides must not be interfered with, and proper culverts must be provided for passing it either under or over the conduit. In connection with the question of lining an open watercourse is that of the velocity of the stream, and it was said that the proposed velocity of neither one nor the other of these aqueducts was such as would be inconsistent with a simple clay bottom; it would not disturb a clay bottom, and would have no effect at all on a gravel bottom; but, nevertheless, an aqueduct for the supply of water to a city should be lined throughout.

Fig. 24, reduced from the sections given in the Appendix A Q of the report, shows the form of the open watercourses of Mr. Bateman's main aqueduct. The fall is 6in. per mile. The width at the surface of the water is 26ft. 4½in., and the depth 10ft., the form being circular throughout. The sectional area of the stream is 193·4 sq. ft., and the wetted sides 35ft. in length. The hydraulic mean depth (H.M.D.) therefore is, $\dfrac{193\cdot 4}{35} = 5\cdot 526$, and $\sqrt{5\cdot 526 \times 1} \times \dfrac{10}{11} = 2\cdot 143$ ft. mean velocity, per second. Then $193\cdot 4 \times 2\cdot 143 \times 540{,}000 = 224{,}000{,}000$ gallons per day carried.

Fig. 25 shows the form of section of the covered watercourses and tunnels. The width is 16ft. 6in. at a height of 3ft. 9in. from the bottom, with a semicircular arch. The depth of the invert is 1ft. 6in., and the depth of water 10ft., the surface of the stream running within a height of 2ft. of the soffit of the arch at the crown. The fall is 14in. per mile, the sectional area 146·25 sq. ft., the wetted sides 34·866ft. H.M.D. $= \dfrac{146\cdot 25}{34\cdot 866} = 4\cdot 195$, and $\sqrt{4\cdot 195 \times 2\cdot 333} \times \dfrac{10}{11} = 2\cdot 844$ ft. per second mean velocity. Then $146\cdot 25 \times 2\cdot 844 = 415\cdot 935$ c. ft. per second, which gives 224,604,900 gallons per day.

During the sitting of this Commission, which extended over thirty-six days, the examination of witnesses took somewhat the form of a discussion, between the members and the principal witnesses, on the question of the quantity of water necessary to be supplied for the growing wants of London, and both Mr. Bateman and Messrs. Hemans and Hassard had to consider whether the aqueducts would carry as much as 300,000,000 gallons per day. It was found that this could be done with but slight additions to the capacity of the aqueducts already proposed. In the case of the Welsh project, an increase of 2ft. in the depth of water in the open watercourses, that is, from 10ft. to 12ft., would accomplish it, and the width at the water surface would then be 27ft. 9in., the fall remaining the same, viz., 6in. per mile. In the covered portions and the tunnels the section would be enlarged, as shown in the Appendix A R of the report, of which Fig. 26 is a reduction.

This section is drawn in Fig. 26 on a larger scale than

the other figures, to show its peculiarities, the chief one of which is the depth of the invert, which is 3ft. If the disposition of the materials be examined, it will be seen that they are placed so as to make the form a very strong one, and at the same time it is one which' affords a great hydraulic mean depth.

There remains a question whether, for an open aqueduct which must for a long time after its construction carry much less water than it is ultimately designed to carry, the circular form of section of the waterbed is as good as one with straight slopes and a nearly straight bottom. When the section of the stream at its greatest flow occupies nearly half the circular bed, the sides are pretty steep; but in the mean time, and while the quantity of water being carried is much less, the sides of the stream are shallow. That is not so in the angular form. At the time these conduits were proposed, the quantity of water supplied to London was about 108 million gallons per day. The open watercourse of Mr. Bateman's main aqueduct is 13ft. 6in. radius, and, as shown above, would run 10ft. deep when supplying 224 million gallons a day, but at the first, and for some time after its construction, the depth corresponding to this quantity would have been only 6ft. 9in. in the centre of the stream.

To carry the quantity supplied to London, taking it at 15,000 cubic feet per minute, the depth would be 7ft. 8in. Thus in about fifteen years the increase in depth would have been 11in.

In the open watercourse of Mr. Hassard's main aqueduct, with the same inclination—viz., 6in. per mile, the depth at the first for the same quantity of water would have been 6ft. 3in. in the middle and 5ft. 9in. at the sides. For 15,000 cubic feet per minute the depth would be 7ft. 2in. in the middle and 6ft. 8in. at the sides. Thus either of these streams would have increased in depth about 11in. between that time and this, but the angular form would from the first have had considerable depth at the sides.

SECTION XV.

RIVERS AND WATERCOURSES.

THE quantity of water flowing down a river channel may be calculated by the following rules: they range, in point of accuracy and attention to details, from the simplest and most general to the nicest refinements of correction for varying conditions of a flow of water; but they all depend essentially upon the two conditions of hydraulic mean depth and inclination of the surface of the water, which together govern the velocity. The volume is the product of the cross sectional area of the stream and the mean velocity of the whole section, and while the former is a matter of simple measurement, the latter is more difficult to arrive at with any degree of accuracy; and it is only after an investigation of the immense number of experiments which have been made, that the authorities in the science of hydraulics have been able to arrive at formulæ which represent the true velocity approximately. With certain precautions, the velocity may, in some cases, be actually measured, as well as the cross sectional area, and when that can be done the volume may be found more satisfactorily in that way. The cross sectional area of the stream may be measured by selecting a part of the river which is of tolerably uniform section for a considerable distance—say 200 yards—dividing the distance into equal lengths, and measuring the cross section of each division by means of ropes—preferably of wire—strung across the river, taking the depths at short intervals along the ropes, and where the difference of level of the surface of the water at the two ends of the length of river experimented upon can be ascertained with accuracy, the mean velocity of the whole stream in ordinary cases and in tolerably uniform channels may be found by the formula deduced by Eytelwein, or that by Du Buât. The fundamental conditions are that the velocity varies as the square root of the hydraulic mean depth, the fall being constant, or as the square root of the fall when the hydraulic mean depth is constant; or, neither being constant, as the square root of the hydraulic mean depth and fall multiplied into each other; and to bring this abstract rule into conformity with observed actual velocities a coefficient is applied, of such magnitude as to bring the abstract numbers into coincidence with the actual numbers observed. Eytelwein's rule for the mean velocity per second is $v = \frac{10}{11}\sqrt{hf}$, where h = the hydraulic mean depth, and f = twice the fall per mile; h, f, and v being all in feet or all in inches.

Du Buât's rule is, for inches,

$$v = \frac{307\,(\sqrt{d} - \cdot 1)}{\sqrt{s} - \text{hyp. log.}\,\sqrt{s} + 1\cdot 6} - \cdot 3\,(\sqrt{d} - \cdot 1)$$

and for feet,

$$v = \frac{88\cdot 51\,(\sqrt{d} - \cdot 03)}{\sqrt{s} - \text{hyp. log.}\,\sqrt{s} + 1\cdot 6} - \cdot 084\,(\sqrt{d} - \cdot 03)$$

d being the hydraulic mean depth, and s the length in which the surface of the water falls one unit, as 2,640 when the fall is 2ft. in a mile. The fractional deductions are made from the fundamental formula—$v = c\sqrt{\frac{d}{s}}$, in which c is the coefficient 307 or 88·51, in order to make it agree more nearly with the results of experiments under varying conditions, and are applicable in cases of moderate velocity of 2ft. or 3ft. per second.

Other authorities make corrections for similar effects of increasing velocity. Thus, Mr. John Neville, in his "Co-efficients and Formulæ," published by Messrs. Crosby Lockwood and Son, gives, from experiments made by various persons, the increasing coefficient as follows, when applied to the abstract rule $\sqrt{rs}$, r being the mean radius or hydraulic mean depth, and s the sine of the angle of inclination, or the fall in any length divided by that length; thus being $\frac{1}{2640}$ when the fall is 2ft. in a mile. For a velocity of about 1ft. per second the coefficient is 91·3; for about 1½ft. per second it increases to 95·5; for 1¾ft. per second, 98·6; for 2ft., 100·5; for 2½ ft., 100·6; for 2¾ft., 103; for 3¾ft., 106·6; for 5ft., 109·3; for 6ft., 111; for 7½ft., 112·3; for 14½ft., 117·9; for 15½ft., 118·4; and for a velocity of about 21ft. per second, the coefficient is 120. As the velocity thus does not strictly follow the rule of $\sqrt{rs}$, Mr. Neville has found a more exact formula, which is — $v = 140 \sqrt{rs} - 11 \sqrt[3]{rs}$, and this seems to agree nearly with observed velocities under all circumstances. Mr. Neville has found that Du Buât's formula may be pretty safely relied on when applied to general practical purposes, and says that much of the valuable information presented by Prony and Eytelwein is but a modification of what Du Buât had previously given, and to whom we are primarily indebted for much that is attributed to the two former.

None of these found any difference in the velocity of a stream which could be attributed to the kind of surface over which it ran; but a later authority, Kutter, has introduced into his formula, as translated by Mr. Lowis D'A. Jackson, a term of correction according to the kind of surface, as brickwork, earth, gravel, &c.

Kutter's formula is, for English feet—

$$v = \left\{ \frac{\frac{1\cdot 811}{N} + 41\cdot 6 + \frac{\cdot 00281}{S}}{1 + \left(41\cdot 6 + \frac{\cdot 00281}{S}\right)\frac{N}{\sqrt{r}}} \right\} \sqrt{rs}$$

in which v = mean velocity in feet per second, r = mean radius or hydraulic mean depth in feet, S = sine of the hydraulic slope of the surface, N = coefficient of roughness and irregularity; and the values given to N are for brickwork and ashlar in good order, ·013; for channels in earth in good average order, ·020; and for rivers and brooks, from ·020 to ·035; but this formula is not so well adapted to rivers as the three preceding.

It may be useful to compare the results of the first three rules. In a river of 100ft. mean width, 6ft. deep, and of such a contour of bed as to give 5ft. hydraulic mean depth, the fall of the surface of the water being 1ft. per mile, or 1 in 5,280, the mean velocity in feet per second by Du Buât's rule is,

$$\frac{88 \cdot 51 \, (\sqrt{5} - \cdot 03)}{\sqrt{2640} - \text{hyp. log.} \, \sqrt{2641 \cdot 6}} - \cdot 034 \, (\sqrt{5} - \cdot 03) = 2 \cdot 67$$

and $2 \cdot 67 \times 600 = 1,602$ cubic feet per second.

By Eytelwein's rule it is, $\frac{10}{11} \sqrt{5 \times 2} = 2 \cdot 87$ and $2 \cdot 87 \times 600 = 1,722$ cubic feet per second.

By Mr. Neville's rule it is,

$$140 \sqrt{5 \times \frac{1}{5280}} - 11 \sqrt[3]{5 \times \frac{1}{5280}} = 3 \cdot 23,$$

and $3 \cdot 23 \times 600 = 1,938$ cubic feet per second.

And if we take the coefficient for velocities about as much as this to be 103, as found by others, the mean velocity would be

$$103 \sqrt{rs} = 103 \sqrt{5 \times \frac{1}{2640}} = 3 \cdot 17 \text{ft. per second,}$$

or, nearly the same as by Mr. Neville's own rule; and it illustrates what is found to be a general tendency in all the hydraulic formulæ derived from experiments necessarily made on a comparatively small scale—viz., that the action of large masses of water is sensibly greater than these formulæ indicate, so that in applying them to rivers of large volume they rather understate the actual quantities, which, indeed, is no fault, but the contrary.

The truth of Eytelwein's and Du Buât's formulæ has been confirmed by Mr. Bateman, for rivers and open watercourses where the section is tolerably uniform. When he laid out the Manchester Waterworks he constructed the works in many parts with special reference to taking such observations as would determine a great many points which were then somewhat in doubt, and he tested upon the watercourses there the calculations of almost everybody. (*Vide* Evidence, Water Supply Commission, 1868.)

But where the fall of the surface of a river is very small in any length that could be experimented upon, as it is in many cases, the inclination cannot be ascertained with sufficient accuracy to enable these formulæ to be applied for finding the velocity, and in those cases the volume of water is best ascertained by actual measurements of the velocity with floats, and for that purpose no better proceeding can be taken than that adopted on the river Thames. The length of the river experimented upon was divided into six measured distances, and the time was taken in which the floats traversed the six divisions at five or six different places in the width of the river. The mean velocity was not computed from the observed surface velocity, but was actually ascertained by floats so adjusted that whilst one all but dragged upon the bottom, and therefore travelled with the bottom velocity, another floated at the surface, and the two floats being tied together, one acted upon the other, in quickening and retarding their respective paces. Gutta-percha was used for the floats; it is very nearly of the same specific gravity as water, being 0·96 of water, and therefore it floats. A gutta-percha ball will float almost wholly immersed, so that the wind can have no effect upon it. The balls were so adjusted that one was heavier than the other, and sank to the bottom, but did not touch it, and thus by finding the mean velocity in all parts of the stream, the mean of the whole stream was found.

To estimate the mean velocity roughly by one observation, Prony's rule may be taken

$$v = \left(\frac{7\cdot783 + V}{10\cdot345 + V}\right) V$$

in which $V =$ the maximum surface velocity in feet per

second in the centre of the river, or in its axis, whether that be in the centre or not, and this gives for velocities similar to those in the above examples, the same results as in Mr. Neville's shorter rule $v = \cdot 835 \text{ V}$, which would indicate a maximum surface velocity of $\frac{4 \cdot 73}{\cdot 835} = 5 \cdot 66\text{ft}$. per second, but for velocities about half these the mean is more nearly $\cdot 8$ V.

Referring to the examples above worked out, it must be confessed that there is no very near agreement between $2 \cdot 67$ as found by the first, and $3 \cdot 23$ by the last; but this velocity is greater than necessarily occurs in rivers, and the best judgment would probably be shown in using Eytelwein's or Du Buât's formula in cases of ordinary flow, and Neville's or Prony's in floods.

As to the velocity of water at the bottom of a river, or anywhere along its bed, it is very difficult to ascertain it by actual measurement, apart from that in other portions of the body, but from the experiments of Du Buât it is found to be, for mean velocities of about 3 ft. per second,
$$u = (\sqrt{V} - 1)^2$$
When V = the mean surface velocity from side to side of the stream, which is always less than that in the centre or axis of the stream. In this case u and V are the velocities in inches per second. Mr. Beardmore adopted this rule in his hydraulic tables for velocities of from 1 to 15ft. per second.

Experiments which have of late been made upon the large American rivers show a different relation between the surface and bottom velocities.

In a discussion at the Institution of Civil Engineers in 1879, Mr. George Higgin said that in the experiments carried out under the direction of Mr. Bateman, Past-President of the Institution, on the great rivers of South America, certain new laws were discovered in the movement of large masses of water. One of these was that the surface velocity of water at a given inclination varied directly as the depth of the channel, and another was that the bottom velocity of water varied directly as the square of the depth. With a given volume of water passing at increasing rates of inclination, and, therefore with diminishing depth, it was calculated that at a depth of 27ft., and with a surface velocity of 176ft. per minute, the bottom velocity would be 69ft. per minute, at a depth of 21ft. and with a surface velocity of 241ft. per minute, the bottom velocity would be 72ft., and at a depth of 18ft., and with a surface velocity of 290ft. per minute, the bottom velocity would be 75ft.; thus while the surface velocity increased 65 per cent., the bottom velocity increased only 10 per cent. But the question arises whether the rules derived from the motion of masses of water so vast as those of the American rivers are more applicable to the English, Irish, Scotch, and Welsh rivers, than the old formulæ are, confirmed as they have been by experiments on a scale of very considerable magnitude.

Investigations by Professor Osborne Reynolds, F.R.S., and Professor W. C. Unwin, F.R.S., have thrown much light upon the motion of water in open channels and in pipes, explaining in a great measure the phenomena observed by the older hydraulicians.

SECTION XVI.

COMPENSATION TO MILLS.

It is hardly possible to divert water from one district to another without injury to the interests of persons already having rights in its use, and where mills driven by water power exist below the point of diversion it is incumbent on the party proposing to divert water for use elsewhere to compensate the millowners for its abstraction. Unless this be proposed to be done in a full and adequate measure, no interference with existing rights will be sanctioned by law. Millowners are very tenacious of their water rights, and sometimes have spent enormous sums of money in opposing waterworks schemes. They have often overrated the value of water power to themselves, not only in corn-mills, where water power is by prejudice presumed to be superior to steam power because of its steadiness, but in woollen and cotton mills, and even in fulling and in rolling mills, where the inequality of the work done during various parts of the day causes a waste of water on the wheels.

To give a money compensation to millowners where the number is considerable, cannot come into question for a moment. Attention must therefore be directed to provide reservoirs to regulate and economise the water, so that the quantity which previously passed the mills in floods and was useless, indeed injurious, may be stored and made use of.

Where mills are compensated from the same ground that supplies the town, it is well to have separate reservoirs for the two purposes, so that the turbid flood waters are made to overleap the channel which conveys the clear water into the reservoir for the supply of the town, and the portion due to the mills to enter that from which they are to be supplied. This may be done by making the two flood weirs at the same level, and proportioning the length of each to their respective quantities, or better by a separating or leaping weir of which this diagram is a sketch.

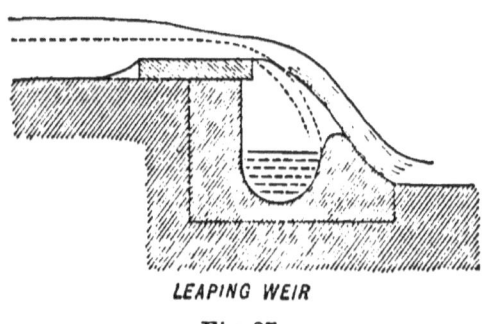

LEAPING WEIR
Fig. 27.

The principle on which the leaping-weir acts is to separate the maximum flow of water required for the town, and to keep it at as high a level as possible, from the greater volume of flood-water, which is not required to be kept at so high a level. The smaller flow is continuous, and, up to the maximum flow required, is comparatively clear, while the flood-waters come down intermittently with a rush. The position of the lip, which separates the clear water from the flood-water, may be fixed on the following principles. For any given depth of water D in the diagram, Fig. 27, let $h = \frac{4}{9} D$, and the parabolic curve $a\ b\ c$ the line of flow of the particles which have the mean velocity of the whole sheet of water. Then, inasmuch as the height h governs this velocity, which would be, say, $7 \cdot 5 \sqrt{h}$ if taken in feet per second, the width W, for any height H, would be $= \sqrt{4\,h\,H}$. If the water arrives at the edge of the weir with any pre-impressed velocity, the head necessary to give it the velocity must be added to that measured as D. If d be this head, then h must be taken $= \frac{4}{9}(D + d)$.

This would be the case where the water runs over a wide-crested weir before arriving at the edge over which it

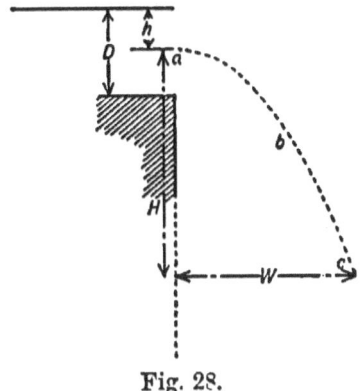

Fig. 28.

falls; and as this velocity of approach increases, so will the width W increase, which would then be

$$= \sqrt{4 \times \frac{4}{9}(D + d)H},$$

d being $= \dfrac{v^2}{64}$ when $v =$ the velocity, in feet per second, of the water approaching the edge of the weir.

Mills generally are so constructed as to be capable of using water at the rate of not more than one-third of the average yield of a district; whether from design or from the long experience of the millwrights of the most economical extent to which water can be applied to mills without large reservoirs, is immaterial, the fact being so; and one-third of the available yield has generally been given to them in compensation, whereby they are abundantly benefited, for they get this third regularly day by day, whereas before they had it very irregularly.

In 1868, as well as in some other years, this allowance has had to be stopped in some cases, in order to furnish a sufficient quantity for the more urgent necessities of the towns, but money compensation has had to be substituted. A reservoir, therefore, constructed solely for the use of the mills, must have a capacity equal to one

K

half that of the reservoir for the town, plus about 5,000,000 cubic feet per 1000 acres of drainage ground, to compensate for its being deprived of the dry weather flow, which, as has been said, is received into the reservoir for the town.

Where it is desirable to take the whole of the water of a district for the supply of the town, whether because of its superior elevation or its purity, and where the mills can be compensated from a reservoir in a separate district, the area to be appropriated to the use of the mills should be of such extent that two-thirds of the yield from it shall be equal to one-third of the area appropriated to the town, because, in addition to one-third of the yield from the latter which is due to the mills, one-third of the waters from whatever area may be appropriated to the mills in the other district is already enjoyed by them; therefore two-thirds of its yield should be equal to one-third of the yield of the district appropriated to the town; or, in other words, the area appropriated to the mills must be one-half as large as that appropriated to the town, the amount of rainfall being the same in both districts; and if the rainfalls be unequal, the compensation area must be more or less than one-half of that appropriated to the town accordingly as the rainfall may be less or more respectively in the one district than in the other.

It may be observed that although a given volume of water falling through a given height produces a certain number of horses power on the basis already stated, viz., that 8·8 cubic feet of water per second falling 1ft. is equal to 1 horse power,—and that, likewise, the same quantity falling 10ft. is equal to 10 horse power—the stream of water exerts its power not for 8 hours a day only, as the horse does, but for the whole 24 hours, and is, therefore, in some cases, of the value of three times as many horses work as its nominal power represents, if the flow of water, while the mill is standing, be stored in a dam close at hand. This, no doubt, causes a certain amount of inconvenience to mills lower down the stream, in respect of the quantity so withheld, which may not come down until a late hour in the day; and on some streams regulations are established whereby a tolerably equable flow is maintained during reasonable working hours; and taking the circumstances of a large number of mills, and comparing the work done with the power which passes them, they appear to do an amount of work, with the assistance of the night water impounded in dams, which is equal to about one-third of the power of the available quantity of water. But with a storage reservoir to control the whole available quantity and deal it out day by day during say 12 or 14 hours each day, the velocity of the stream is increased, and the mills get the water at earlier hours all the way down. The benefit to be derived from the construction of a storage reservoir is in respect of two-thirds of the available quantity of water. Thus, with a storage reservoir, about three times the quantity of work may be done, but the benefit derived is in respect of twice the quantity only.

The following is an instance in point. There are fifteen mills in a district above which there is a suitable site for a reservoir which commands a watershed area of 1,000 acres or thereabouts. The average annual rainfall for 14 years has been 33 inches. The loss by evaporation and absorption has not been ascertained in this particular area, but comparing its position and general circumstances with other districts in which actual measurements have been made, it may be estimated with great probability at 13in. We say we lose these 13in. by evaporation and absorption, but it means the whole loss from whatever cause. Of course a good deal must be evaporated, and another portion absorbed by vegetation—this however being mostly evaporated from the leaves—and by sinking into the ground, from which it does not again issue within the watershed area belonging to the reservoir; but also it is very probable that over a large area the rainfall is not uniform, and that parts of it will receive more rain than other parts, and if the rain gauge be situated in one of these wetter portions of the district, it will show a result greater than the true one. This is more especially likely to be the case where the rainfall is estimated not from a gauge planted within the district itself from which the water is to be derived, but from one at some distance from it. So that when it is said that so much is lost by evaporation, it is meant to say that the whole quantity due to the watershed area and the rainfall as shewn by the gauge does not come into the reservoir. The loss varies from 10 to 18 inches as determined by different gaugings, and a part of this great variation may perhaps be due to the situation of the rain-gauge with reference to the actual watershed area to which its register is applied.

The question may arise, how it is known that either 10, 13, or 18in. of rainfall are evaporated and absorbed? The result is arrived at by labour and patience in gauging the quantity of water actually flowing off the ground, by which it is found that the actual quantity falls short of the whole quantity due to the rainfall by certain quantities which vary between those limits, and comparing these various quantities with the character of the ground upon which the rain has fallen in each case, engineers exercise their judgment as to what allowance to make in any particular district. If 33in., as in this case, be the average annual rainfall, and 13in. the loss there would be 20in. available if all of it could be stored, but as that is not practicable, a deduction of about 5 inches should be made for excessive floods, leaving a practically available depth of 15in. A depth of 15in. over 1,000 acres would give the following quantity:—There are in 1,000 acres 43,560,000 square feet, which at 15in. deep would yield 54,450,000 cubic feet. If we take 312 working days in a year, and supply the mills with water for 12 hours a day, we should be able to give out of this reservoir year by year $\left(\dfrac{54{,}450{,}000}{312 \times 12}=\right)$ 14,540 cubic feet per hour, or 242 cubic feet per minute. But as the mills are supposed to be capable of beneficially using one-third of this already, the benefit to be derived would be in respect of 160 cubic feet per minute only. Will it, then, pay to make this reservoir to give out this quantity for 12 hours a day all the year round, and year by year?

Seeing what many other reservoirs have cost, and not going to either extreme, this one might cost £12,000, the bank being made as perfectly as possible, by raising it in thin layers with barrows and dobbin carts, the puddle being particularly well worked, the waste weir of good length, so that on the occurrence of a rare flood, when the reservoir might be nearly full, the overflow could not attain to such a height as to endanger the bank by over-topping it, a bye channel being provided for the ordinary flow of the stream, and a substantial cottage built for the reservoir keeper. The falls at these 15 mills vary from 10ft. to 30ft., the aggregate being 326ft. The wheels are all either overshot or breast wheels, of good construction, and in good working order. Then, although we estimated 13·2 cubic feet of water per second per foot of fall to be, perhaps, an average of all wheels to give an effective horse-power, we shall be warranted in taking 12 cubic feet per second in this case. This is the quantity that Sir Wm. Fairbairn frequently gave as the proper quantity in such cases. The aggregate amount of power, therefore, to accrue to the mills from the construction of this reservoir will be $\left(\dfrac{160 \times 326}{12 \times 60}=\right)$ 72 horse-power. In the district we are considering an effective horse-power is worth £12 per annum, and if each mill contribute its quota according to the benefit it would receive (we will not now go into the more minute subdivision of benefits by considering that those mills nearest the reservoir would receive more benefit than those farther off) there will be an income of £864 a year. Deducting the wages of the reservoir keeper, and setting aside £40 a year to provide a fund for repairs, there would be a net income £720 a year. The outlay being £12,000, the percentage return would be 6 per annum. But this does not represent the whole gain to the mill. The gain is chiefly in having a constant stream of water, and an absence of floods. Nothing is more annoying to a miller than to see water pass him which he can make no use of, and which at the same time hinders his work by backing up against the tail of his wheel. By storing the flood waters he can always reckon upon a certain power of doing work, and can enter into contracts with a certainty of being able to fulfil them. Herein lies the chief gain in constructing storage reservoirs.

SECTION XVII.

Of Water Power in General.

A CUBIC foot of water, or $6\frac{1}{4}$ gallons, weighs $62\frac{1}{2}$lb. The gallon weighs 10 lb. It is commonly taken from Watt's experiments that a horse of average power does work equivalent to that of raising 33,000lb. 1ft. high per minute, or 550lb. 1ft. high per second, if not worked more than 8 hours a day. This would be equivalent to raising 8·8 cubic feet of water 1ft. high per second, or 528 cubic feet per minute. But that quantity of water could not practically be raised to that height in that time by that power. To put everything into the simplest form, the horse might draw water out of a well by means of a barrel attached to a rope passing over a pulley at the head of the well, the horse walking away from it at the speed of 220ft. per minute, or at the rate of $2\frac{1}{2}$ miles an hour, which is taken to be the best working pace during 8 hours a day; and the horse would raise at this speed a weight of 150lb. But the weight which would be thus drawn out of the well would include that of the barrel and the rope, and the hindrance caused by the friction of the pulley on its bearings. The friction of the pulley would be so small that it might be neglected, but the weight of the barrel and the rope would probably be 20 per cent. of the whole weight. The weight of the water raised would be 80 per cent. of the 150lb., or 120lb., which would be 12 gallons. But it would not be an economical way of applying the horse's power, that the load should thus be drawn up by a straight run away from the well, because the quantity of water is small, and time would be wasted in emptying the barrel and replacing it so often.

A more economical arrangement would be to wind the rope round a horizontal drum at a little distance from the well, supported on a vertical shaft, and to attach to the shaft a beam or pole, the end of which the horse could draw in a continuous round, and if the circumference of the path be made 5 times the circumference of the drum the horse would move 5 times as much weight with one-fifth as much speed: that is, 750lb. at the speed of 44ft. per minute. The weight of the barrel and rope in this case would not be more than about 15 per cent. of the whole load, but the resistance of the drum to being turned would be about 5 per cent., so that the whole resistance would be the same, viz., $750 \div 5 = 150$lb. at the speed with which the horse moves; and $150 \times 220 = 33,000$lb. per foot high per minute = 1 horse-power. The actual weight of water raised would be 80 per cent. of the 750lb., or 600lb., which would be 60 gallons each turn. If the well is large enough to allow two barrels to pass up and down, and the end of the horse beam be fitted with a swivel, so that at the end of the motion in one direction the horse can reverse the motion and draw up another full barrel while the empty one is being lowered, the 15 per cent. loss due to the weight of the barrel and rope would be most of it saved, but about as much would be lost in the time required by the horse to reverse the motion at the end of each draw, so that in either case it comes to about the same thing; and in this way the horse can work 8 hours a day. If a horse has a single run out and returns unloaded, the time of working might be 10 hours a day, but the amount of work done in the day would be no more than in the other case.

A horse-power then being 33,000lb. raised 1ft. high per minute, or its equivalent, 33,000lb. falling 1ft. per minute, or 550lb. falling 1ft. per second, the volume of water falling 1ft. is $550 \div 62 \cdot 5 = 8 \cdot 8$ cubic feet per second, to produce 1 horse-power. There are few, if any, mechanical means by which more than 80 per cent. of this power can be transmitted to the working point, or by which more than 7 cubic feet of water per second can be raised back again to the height of 1ft. by the fall of 8·8 cubic feet through that height in that time.

When a river or other stream of water flows in the condition called *in train*, the water can do no more work than it is doing; it gravitates, indeed, with the same power per foot of its fall as if it fell over a precipice; but while it is flowing in train, the whole power with which it is endowed by the force of gravity is expended in overcoming the resistance of the bed to its transfer from place to place, with the rate of motion which it acquires by virtue of the inclination of its volume; but if, at the end of this inclination it fall vertically, it is no longer in train, there being no longer the resistance of the bed, and the force of gravity accelerates its motion, acceleration before being prevented by the continued resistance of the bed, the motion of the water being thereby made uniform, or at least not accelerated. When the same water falls vertically it may be again put in train by interposing a resistance such as will prevent the acceleration of its velocity, and its power may be developed and used by transferring the accelerating force to a uniformly-moving body, the weight of the water and the height of its descent being the measure of its power. The resistance of the bed of the channel being suddenly removed, and that of the unloaded wheel or turbine substituted, the measure of useful effect of the water is the difference between these two resistances per foot of fall in each case. In the latter case a portion—the greater portion—of the power is liberated. Wheels and turbines, according to the perfection of their construction, take more or less power to turn them before they are brought to the point of doing useful work, and this is so much taken from the power of the water, the remainder only being the useful effect of the motor. To compare the power of one force with another, the assumed power of a horse is established at 33,000 foot-pounds per minute, however that may vary from the actual power of different horses. A horse-power is that resistance—viz., 150lb., which custom has established that a horse can overcome when moving along a level road at the speed of 220ft. per minute, during eight hours a day; and $150 \times 220 = 33,000$ foot-pounds per minute, including all friction and resistance of the vehicle or machinery by which the load is moved. It is probably more than most horses can do; but that is of little or no importance, the object being to establish a standard power by which one force may be compared with another; and as to the duration of eight hours a day, that is essential in the question of the power of a horse, but not of a horse-power.

SECTION XVIII.

Water-wheels.

As a cubic foot of water weighs 62¼lb., a quantity equal to 528 cubic feet per minute is equal to a horse-power per foot of fall of the water, when the motor intervening between the power and the work is acted upon by the steady pressure of the water, as when it is delivered into buckets of such form as to lower it down gradually from the upper level to the lower one. In unloaded motors, driven by a steady pressure with the speed required when doing work, the quantities of water required to move them vary with their construction between ⅛th and ⅔ths of the power of the water. The usual portion of the power of the water liberated by such a wheel is about two-thirds, the remainder being absorbed in moving the wheel itself, and the gearing through which it moves the machinery.

When the stream of water is constant in volume, the construction of the motor can be adapted so as to derive from it 80 per cent. of its power, and transfer that portion to the work to be done—to the resistance to be overcome; but when the volume of water varies much from time to time, the motor works at a disadvantage with the smaller quantities, for the strength and consequent weight of the motor, must be sufficient for the largest volume, and the power required to drive it when unloaded bears a larger proportion to the whole power than when the larger quantities of water are being used. When the construction of the wheel or other motor is such that it yields two-thirds of the power expended upon it, half as much more water is required to produce an effective horse-power upon the work done—that is, 792 cubic feet per minute per foot of fall. It is not unusual to reckon that 12 cubic feet of water per second, or 720 cubic feet per minute, falling upon a well-constructed bucket-wheel, is equal to an effective horse-power per foot of fall. This is at the rate of 73 per cent. of the power expended.

The buckets are best made of sheet-iron, where the water is not of such character as to cause an unusual amount of corrosion or deterioration, in the form, or some similar form, to that shown in Fig. 29 ; but they may be made of wood, which has, in some situations, an advantage over iron (Fig. 30). The width of the entrance B C in

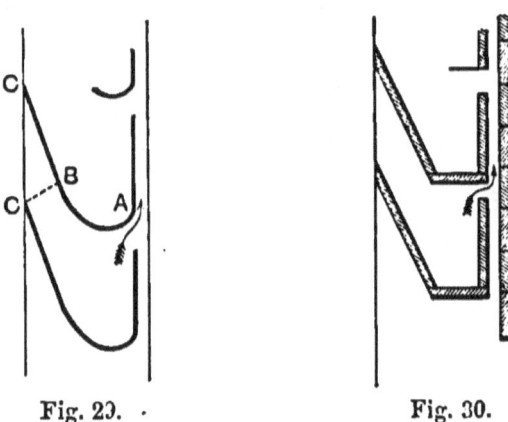

Fig. 29.   Fig. 30.

Fig. 29 is about a third of the depth of shrouding A C, which is about the same as the distance of the buckets apart, C C. This varies from 1ft. to 1½ft., which determines the number of buckets, according to the diameter of the wheel.

The term "overshot" is usually applied to a wheel which receives the water near the top. Formerly the water was carried over the wheel and delivered upon it on the opposite side to that from which it approached, the wheel thus at the top revolving in the same direction as the stream of water, and at the bottom against it ; thus the water overshot the wheel; but the back-lash of the water in the stream below caused a serious hindrance to the motion of the wheel, and the form was therefore changed to the pitch-back, in which the water is delivered upon the wheel on that side from which it approaches it, and consequently at the bottom revolves in the same direction as the stream passing under it. The old term, however, is still retained for wheels upon which the water is delivered near the top, to distinguish them from those which receive the water nearer the centre, called high breast, or low breast, according as they receive the water above or below the centre.

High-breast and overshot wheels may be classed together as being wheels with buckets, and having about the same degree of efficiency, while those which receive the water below the centre have straight float-boards to which the action of the water is confined by a close-fitting breast of masonry. A high-breast wheel has the water laid on at a point not exceeding about 30° above the horizontal centre; above that it may be called overshot. A water-wheel may be likened to a clock-face; it is an eleven o'clock wheel when the water is laid on at 30° from the vertical centre, half-past ten o'clock when laid on at 45°, and a ten o'clock wheel when laid on at 60° below the vertex. The farther from the vertical centre the bulk of the water acts upon the wheel the greater must be its effect, for it has then the greater leverage, and less dead-weight upon the bearings; but then the size and weight of the wheel must be increased, so that the most economical point of application has to be sought between these two conditions, in each particular case in practice.

In every case a small portion of the head of water is taken up in giving a sufficient velocity to the water entering the buckets to fill them to the required degree —generally about two-thirds full—and to prevent the buckets striking against the stream of water; therefore, the entering water should have a velocity rather greater than that of the circumference of the wheel.

This varies from 3ft. to 6ft. per second; 3ft. or 3½ft. is the best for effective power; but sometimes the arrangement of the machinery makes it advisable to give to the circumference of the wheel a velocity as much as 6ft. per second, or even a little more in some cases; and up to 6ft. the increased velocity is not found in practice to much diminish the effective power; but perhaps the best velocity on the whole is 4ft. or 4½ft. per second. The co-efficient of the velocity of the entering water would, in most cases, be properly taken at ·94, in which case the head of water to produce a velocity of 6ft. per second would be $\dfrac{6^2}{64 \times (·94)^2}$ = ·64 ft., or 8in., and allowing for the construction, the loss of head would be about 1ft. The effect of the separate quantities of water lying in the buckets round the wheel between the point of application and that of the discharge near the bottom is the same as the quantity contained in each bucket multiplied into the sum of the horizontal distances of all the buckets from the centre of the wheel. As the velocity of the circumference of the wheel is supposed to be uniform, all the buckets will be filled to the same degree, the sluice-opening and the head remaining the same; therefore, the water may be supposed to form a continuous ring round that part of the circumference of the wheel upon which the water acts, as between A and B in Fig. 31; A being the point of application, and B the point of discharge; and the effect of the weight of water in all the buckets is the same as that of a continuous ring of the same weight multiplied into the distance C of the centre of gravity of the ring from the centre of the wheel, which is, according to the rule made and provided in such case, that the distance of the centre of gravity of a circular arc from its centre is a fourth proportional to the length of the arc, the radius, and the

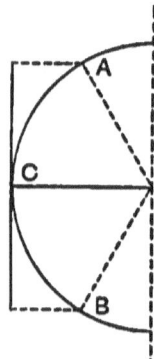

Fig. 31.

chord of the arc. If, for instance, the water be laid on at "11 o'clock," or at 30° from the vertex, and be retained in the buckets until it arrives within 30° of the vertical through the axis of the wheel, the arc will be 120°, or two-thirds of the semi-circumference of the wheel. If R be the radius, the length of the arc will be $\frac{2 \times 3 \cdot 1416 \text{ R}}{3} = 2 \cdot 0944$ R.

The chord is 2 cosin 30° = 2 × ·866 R = 1·732 R; and according to the rule the distance of the centre of gravity of the ring of water from the centre of the wheel is 2·0944 R : R : : 1·732 R : the distance required $= \frac{1 \cdot 732 \text{ R}^2}{2 \cdot 0944 \text{ R}} =$ ·82 R. If the radius be 15ft. to the centre of gravity of the water in the buckets, the whole weight of water would act at a distance of 15 × ·82 = 12·3ft. If the buckets be of such dimensions, and be filled to such degree that the quantity of water poured into them would form a continuous ring 6in. deep, the radius 15ft. passing through the centre of it; then the length of the ring of water would be 2·0944 × 15 = 31·41ft., and the quantity of water would be 31·41 × ·5 = 15·70 cubit feet per foot in width of the wheel, acting at the distance 12·30ft. from the centre. This is, in effect, the same as a vertical column of the height A B, and of the same sectional area as the ring, acting at the distance 15ft. The height of the column is 1·732 R = 1·732 × 15 = 25·98, or say, 26ft.

Let the quantity of water to be expended be 20 cubic feet per second, or 1,200 cubic feet per minute. If the velocity of the wheel at the centre of gravity of the water in the buckets be 5ft. per second, the quantity would be 2½ cubic feet per second per foot in width of the wheel. The width, therefore, would require to be $\frac{20}{2 \cdot 5} =$ 8ft. Such a wheel might be made to give at the working point from 75 per cent. to 80 per cent. of the power of the water. This power is $\frac{1,200 \times 26}{528} = 59$ horse power, and the effective power would be, probably, 45 horse power, and the wheel would allow of the power being increased to 60 horse power at times when one-third more water than usual might be at command; that is, when it would be sufficient to fill the buckets to the degree which would be equal to a continuous ring 8in. deep, 8ft. wide, and having a clear vertical fall of 26ft.

UNDERSHOT WHEELS.

The undershot wheel is an improved form of current wheel; both depend for their efficiency on the impulse of water, as distinguished from its gravity. But instead of being placed in an open current, the wheel is fixed in a close channel, or race, of masonry, very little wider than

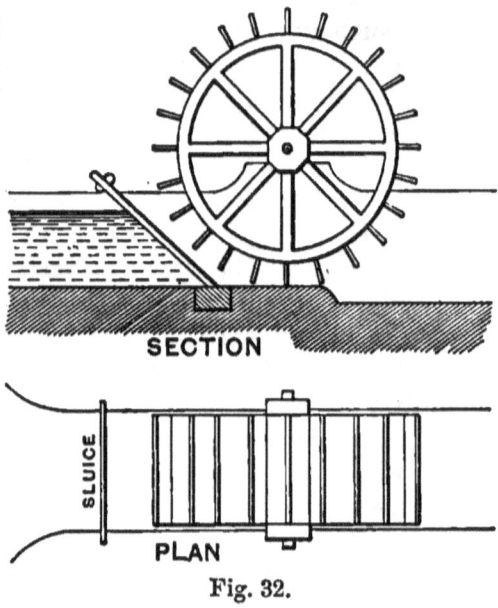

Fig. 32.

the wheel itself, and the water is penned up by a sluice-gate, moving in grooves in the side walls, issuing beneath it when raised by means of a rack and pinion. Fig. 32 represents such a wheel. Because the efficiency of this form of wheel depends upon the velocity which can be induced in the jet of water, the approach to the opening through which it issues is differently formed from other openings through which merely quantity of water is required; the side walls being made to converge towards the opening, and the sluice-gate being placed in a sloping position; at least, this should be the form. Sometimes the sluice-gate is placed vertically; but the effect of placing it in a sloping direction is to increase the velocity ef the water issuing from under it, as compared with that through an abrupt opening, and as the effect of water acting on an undershot wheel is in respect of its velocity —and actually as the square of its velocity—any increase of that element which can be obtained by arrangement of sluices is to be desired. Water issues from openings into the air with the velocity due to heavy bodies falling from a height equal to the head of water, or the vertical height from the centre of the opening to the surface of the pent-up water (with however some abatement, to be hereafter referred to), and when the opening is so formed as not to impede the flow of water, the quantity issuing is nearly that due to the area of the opening multiplied into this velocity.

Many openings from which water is made to issue are so abruptly formed that the actual quantity is much less than the quantity so found, because, although the water attains the velocity due to its head in the axis of the vein, or jet, at some short distance outside the opening—that is, at the point of contraction of the vein—yet it does not do so equally throughout its cross-section, the friction against the sides of the opening retarding the flow of the outer portions. When water flows out of an opening into the atmosphere it approaches it from all sides, as shown in Fig. 33, and, in issuing, its viscidity causes its particles to shoot across towards and meet in the section of contraction (A A in the figure). The cross-sectional area of this part of the jet cannot be measured in practical works, but on a smaller scale it has been

L

measured and compared with the area of the opening. Experiments made on the flow of water through holes in thin plates—and, therefore, under circumstances where

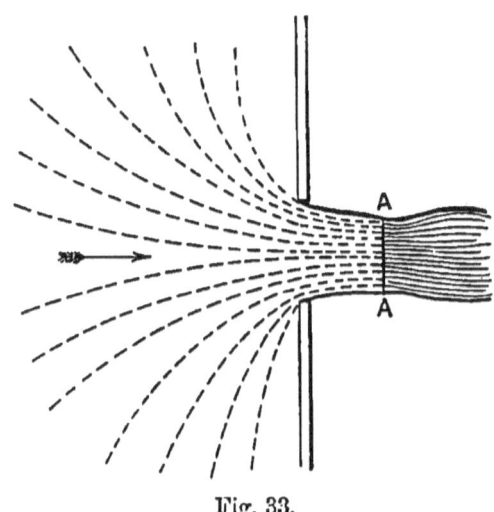

Fig. 33.

the water issues abruptly—show the sectional area of the *vena contracta* to vary from about one-half to about two-thirds of the area of the orifice from which the water issues.

In constructing the sluices of undershot water-wheels they should therefore be made so that the water may have an easy approach to the opening through which it is to issue, for then the retarding influence of the edges of an abrupt opening being reduced, the mean velocity of all the filaments of water passing through an opening will be much increased, if the opening be placed in such a position as to occupy the place of the contracted vein.

This question of increased velocity must always be taken in conjunction with the area of the opening, for it is not an absolute increase, but a relative one—relative, that is, to the area of the actual opening. If an abrupt opening be placed at a distance in front of the point where the issuing jet is to take effect, equal to the length

FLOW FROM ABRUPT OPENINGS. 147

of the contracted vein, the velocity at the point of contraction would be the same as through an opening with converging sides; practically the same, although there are minor influences which affect it slightly. It is on these considerations also that the mouth of a pipe into which water enters from a reservoir is made bell-mouthed, or trumpet shaped, so that immediately after entry the pipe may run full bore.

The damage done by floods in Italy some years ago might lead one to suppose that in that and neighbouring countries hydraulic laws are not understood. Sir John Rennie, in a letter to the *Times*, showed how floods might there be prevented, or at least how damage from them might be obviated; and yet it is chiefly from continental experimentalists that English engineers have received their elementary knowledge of hydraulics.

On that part of the subject now under notice—the phenomena of spouting fluids—the following are the original authorities, and the results of their experiments. Du Buat (*Principes d'Hydraulique*) gives the relation between the diameter of the jet at the point of contraction and that of the orifice as 6 to 9, from which are deduced the relative areas 4 to 9; the Abbé Bossut (*Hydrodynamique*) as 41 to 50, or in area as 16·8 to 25; Daniel Bernoulli (*Hydrodynamique*) as 5 to 7, or in area as 25 to 49; J. B. Venturi (of Modena) as 4 to 5, or in area as sixteen to 25; Michelotti (of Turin) as 4 to 5, or in area as 16 to 25; MM. Poncelet and Lebros, French engineers, from a number of experiments on a larger scale, found the relative areas to be as ·62 to 1, or nearly as 5 to 8. It will be found useful to compare these latter relative areas with the co-efficients of actual discharge to be presently mentioned.

The first practical step to be taken in constructing any work in which water is to be used for power must be to ascertain what quantity of it will issue from any opening of given dimensions under a given head of water, and, therefore, it is worth while to go to the root of the matter before proceeding to particulars of construction. Although the experiments made on the form which water assumes when issuing from an opening, and to which we have before referred, have necessarily been made on a smaller scale than that of general practice, yet they are very useful guides to the comprehension of the reasoning on larger affairs, and help to account for anomalies between that which we find to be the result by experiment, and that which we should expect on theoretical grounds alone. But it is always so in constructive works; the best and only true results are to be arrived at by comparing theoretical knowledge with experimental knowledge, and ultimately uniting them in practice.

Experiments on the actual quantities of water issuing from openings of known areas and forms show that those quantities are in all cases less than those which are due to the areas of the openings multiplied into the velocities due to the several heads of water, calculated upon the known effects of the power of gravity, which give in any particular case the quantity or the velocity which is called the theoretical quantity or the theoretical velocity. But taking the velocity so found in feet per second—viz., eight times the square root of the head of water in feet, as a standard, co-efficients of diminution have been found for various forms of opening, which show the ratio between the theoretical velocity and that which may be called the mean velocity of all the filaments of water issuing through any actual opening. The more abrupt the outlet the less is the co-efficient of actual velocity, or, as some say, the co-efficient of contraction. Whether one or the other of these expressions is the more proper depends upon the form of the opening, the area of which is measured and multiplied into the co-efficient to find the actual discharge. With abrupt openings the diminution may more properly be attributed to the contraction of the vein outside the opening, and with openings which have converging side walls the diminution should be attributed to the reduction of velocity caused by the friction of the particles of water amongst themselves and against the side walls. Generally, the velocity is found by measuring the actual quantity of water discharged in a given time, and dividing it by the area of the opening through which it issues. In making these experiments the water is discharged into a tank of known capacity, and the time of filling is accurately noted. From experiments conducted in this manner co-efficients are found which make theoretical deductions accord with observations of facts.

There is another method of ascertaining the velocity by actual measurement. It is known that a heavy body issuing into the atmosphere with an initial velocity at some height above the ground describes in its descent a parabolic curve nearly, and would do so exactly if the air were removed. If an experimental vessel with a hole in its side near the bottom be set up at some height above the ground, and if a vertical line be let fall from the mouth of the orifice to a horizontal plane at any depth below it, and if horizontal lines be measured out from the vertical line to the jet at several points, then these ordinates, measured at the points of corresponding abscissæ, show the jet to describe a parabolic curve nearly. If h be put for the head of water above the centre of the orifice, which is due according to the theory of falling bodies to the initial velocity of the issuing jet, and if x = the length of any abscissa and y its corresponding ordinate, then $y^2 = 4hx$. By measurements in this way Bossut found at three different heights the value of h to be nearly that of the actual head of water. Calling the actual head H, the heads calculated from the range of jet at these several heights were as follow:—

When H = 7·511ft., h = 7·404ft.
 „   H = 12·890ft., h = 12·564ft.
 „   H = 23·583ft., h = 22·720ft.

The differences are due partly to the resistance of the air at high velocities, and partly to the viscidity of the water and consequent friction of its particles.

If we state the theoretical discharge or the theoretical velocity as = 1, the limits between which actual discharges and actual velocities vary are ·625 and 1 of those determined by the theory of falling bodies. By no form of construction can the actual discharge or the actual velocity reach the theoretical discharge or the theoretical velocity, because the theory assumes the falling body to fall *in vacuo*, and although the very great difference between the density of the atmosphere and any heavy body falling in it, or issuing into it, reduces its retarding influence to a very small opposing force, yet it has a degree of density, and its opposing force is inversely proportionate to the difference of density between itself and the body issuing into it. The quantity of water, therefore, or its velocity, issuing from any opening of whatever form, can never quite reach that found by theoretical calculation.

Dr. Downing, Professor of Civil Engineering in the University of Dublin, in his 'Practical Hydraulics,' gives a table of experiments — apparently those made by Castel and D'Aubuisson — to determine the coefficient of discharge and that of velocity through short-truncated tubes, the sides of which converge at various angles. The length of the experimental tube was $2\frac{1}{2}$ times the diameter at the outer end. See Fig. 34.

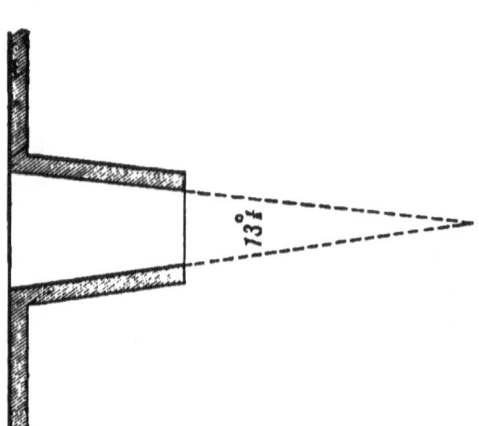

Fig. 34.

The angle that gave the maximum discharge was between 13° and 14°, the ratio between the actual and the theoretical discharge being then as ·95 to 1, but it does not vary much on either side of this ratio within a range of about 5°; thus at 8° it is ·93, and the same at 18°, so that for quickness of discharge the angle may vary from 8° to 18°, with a mean co-efficient of ·94 or thereabouts; but as the angle increases beyond about 18° the co-efficient of discharge diminishes more rapidly, until at 48° it is but ·84, and if continued to the extremest angle of 180° would be about ·5. The co-efficient of *velocity*, however, continues to increase with the angle, even beyond the angle of maximum discharge, where it is ·96, to 48°, where these experiments show it to be ·98. These experiments were made on the small scale, but three experiments made at a mill on the canal of Languedoc show similar results, as to discharge, through a truncated pyramidal tube of considerable dimensions. The length was 9·59ft., dimensions at larger end 3·2ft. by 2·4ft., at the smaller end ·623ft. by ·443ft. The opposite faces made angles of 11° 38' and 15° 18'. The head of water was 9·69ft. The co-efficients of discharge deduced from these three experiments on the large scale were ·987, ·976, and ·979 respectively, or very nearly the full theoretical quantity.

Eytelwein ('Handbuch der Mechanik und der Hydraulik'), translated by Dr. Thomas Young, and quoted in Tredgold's 'Tracts on Hydraulics,' gives the constant multiplier 7·8 for an orifice of the form of the contracted stream, that is when the internal edges of an orifice in a thick plank or a short tube are rounded off to that form, from which we deduce the co-efficient ·97. When the internal edges are not rounded the co-efficient is ·92.

The late Mr. Beardmore, in his 'Tables to Facilitate Hydraulic and other Calculations,' gives the constant 7·5 to be multiplied into the square root of the head of water, for the mean velocity in feet per second for openings such as these. The theoretical value is $8·03\sqrt{h}$, so that the co-efficient in this case would be ·934.

From the results of his own practice, which was considerable in this respect, Mr. Beardmore considered that the constant 7·5 was to be preferred for openings constructed as nearly as may be of the form of the issuing jet.

Adopting this constant, we may apply it to an example. If an undershot wheel be applied to utilise a fall of water of 4ft.—that is 4ft. head from the surface down to the centre of the opening of the sluice; and if the width of the sluice be 6ft., and the vertical height to which it is found necessary to raise the sluice-gate, in order to do the work required, be 6in.; the area of the opening would be 3 sq. ft., which call A; let h = the head of water in feet, then the mean velocity of all the filaments of water through that opening, supposing the side walls to be splayed, and the head of the sluice-gate sloped up-stream, would be $7 \cdot 5 \sqrt{h}$, and the quantity of water discharged would be $7 \cdot 5 \sqrt{h} \times A = 7 \cdot 5 \times 2 \times 3 = 45$ cubic feet per second.

As to the best diameter of an undershot wheel and its speed, Sir W. Fairbairn states, in his 'Treatise on Mill Work,' that assuming $2 \cdot 4 \sqrt{h}$ to be the velocity of the extremity of the float-boards to give a maximum effect, and let N = the number of revolutions per minute, then the diameter, expressed in terms of the velocity and height of fall, will be $19 \cdot 1 \times \dfrac{2 \cdot 4 \sqrt{h}}{N} = \dfrac{46 \sqrt{h}}{N}$ nearly. For instance, if the fall be 4ft., and the number of revolutions 8 per minute, then the diameter $= \dfrac{46}{8} \times \sqrt{4} = 11\frac{1}{2}$ft. nearly.

The number 19·1 appears to be arrived at thus: N, the number of revolutions, is taken in respect of a minute, while the velocity $2 \cdot 4 \sqrt{h}$ is taken at per second. The number of seconds in a minute (60) divided by 3·1416, the ratio of the circumference of a circle to its diameter, = 19·1.

Other authorities give the best speed of an undershot wheel as half that of the water; others again as ·57 of that the water. Now there is an apparent discrepancy here, and it probably arises from the different values given to the co-efficient of velocity of the issuing jet. According to the form of the opening, that velocity varies from $5 \sqrt{h}$ to $7 \cdot 5 \sqrt{h}$, being least with abrupt openings and greatest with openings of the form of the *vena contracta*, or with trained walls and sluices. If the co-efficient 5 be adopted, and the proper speed of the wheel be taken at half the velocity of the water, then the speed of the wheel would be $2 \cdot 5 \sqrt{h}$, agreeing nearly with Sir W. Fairbairn's statement above given; but with openings of the form to which it would be proper to apply the co-efficient $7 \cdot 5$, the speed of the wheel (supposing it be likewise half that of the water) would be $3 \cdot 75 \sqrt{h}$. On the whole, it will probably be near the truth for ordinary cases to make the speed of the wheel in feet per second from $3 \sqrt{h}$, to $3 \cdot 5 \sqrt{h}$, accordingly as the form of the opening varies either way from a form which would be a mean between an abrupt opening and one of the best form; the co-efficient, or rather constant multiplier, being taken at from 6 to 7 for the velocity of the water, and the speed of the wheel at half that velocity.

The ratio of the useful effect, or work done, to the power expended on an undershot wheel is usually reckoned as but little more than 1 : 3. Thus, if the head of water be 4ft., the width of the opening 8ft., and its vertical height 6in., the area would be 4 sq. ft. The velocity of the water through the opening (supposing there to be converging side walls and sloping gate) would be $7 \cdot 5 \sqrt{h} = 7 \cdot 5 \times 2 = 15$ft. per second, and the area being 4 sq. ft. there would be 60 cubic feet of water expended per second, or 3,600 cubic feet per minute, falling 4ft., which is the same power as $(3,600 \times 4 =)$ 14,400 cubic feet falling 1ft. in a minute. But as the ratio of the useful effect and the power is as 1 : 3 the effective force of this expenditure of water is $\left(\dfrac{14,400}{3} = \right)$ 4,800 cubic feet raised 1ft. high per minute. A horse-power being 33,000lb., raised 1ft. high per minute is also equal to $\left(\dfrac{33{,}000}{62 \cdot 5} =\right)$ 528 cubic feet of water raised 1ft. high per minute. Therefore the effective power of an undershot wheel applied to utilise this quantity of water with this fall would be $\dfrac{4{,}800}{528}$ = 9 horse-power.

Sir William Fairbairn states that the usual range of diameters for undershot wheels is from 12ft. to 25ft., and that from 12ft. to 16ft. is considered to be most effective. In his own practice he found from 14ft. to 18ft. diameter to give the best duty. The float-boards, instead of being set in a radial direction on the wheel, are sometimes set in an inclined direction, but this does not seem to increase the useful effect. The number of floats is usually $\dfrac{4d}{3} + 12$, d being the diameter of the wheel in feet. Thus, for a wheel 15ft. diameter the number of floats = $\dfrac{4 \times 15}{3} + 12 = 32$. The thickness of the vein of fluid may be from 6in. to 9in., and the depth of the float-boards from 18in. to 24in.

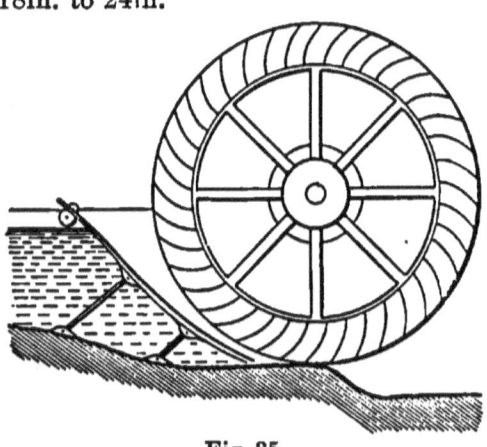

Fig. 35.

The greatest improvement on the undershot wheel has been made by M. Poncelet. His form of wheel is shown in Fig. 35. It has curved floats of sheet iron; indeed, the whole framework is of iron, and very light. It is well suited for low falls and large quantities of water, and can be driven at a greater speed than a low-breast wheel, with which its performance is to be compared. Its useful effect is from 50 to 60 per cent. of the power of the water expended, and its speed from half to three-fifths of the velocity of the water. M. de Bergue exhibited at the Institution of Civil Engineers a drawing of one of these wheels erected by him 16ft. 8in. diameter and very wide, the fall of water being 6ft. 6in. The following is a method of describing the curve of the floats of M. Poncelet's wheel. Let $a\,b\,c$, Fig. 36, be a horizontal line, and $b\,d$ the perpendicular to it. Let $b\,r$ be the radius of the wheel, the divergence from the vertical being from

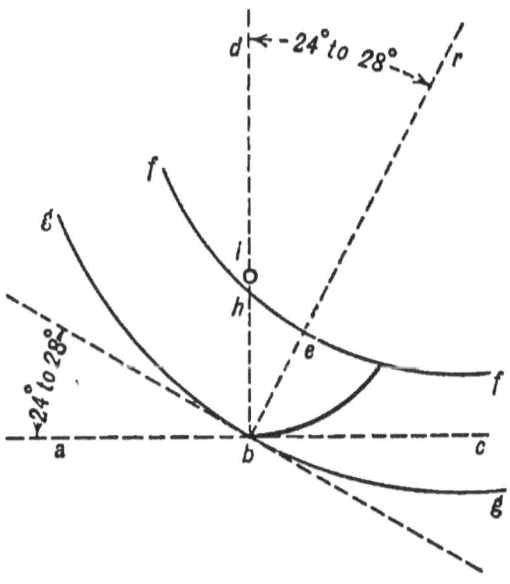

Fig. 36.

24° to 28°. Take $b\,e = $ from one-third to a quarter of the fall of water, and draw the inner circumference ff, and the outer circumference $g\,g$. Take on $b\,d$, $h\,i = $ one-sixth of $b\,h$, and from the centre i, with radius $i\,b$, describe the curve of the float. The number of floats is greater in this than in ordinary undershot wheels, and therefore the motion is more equable and continuous. Let $d = $ the diameter of the wheel in feet, then for wheels from 10ft. to 20ft. diameter, the number of floats $= \dfrac{8\,d}{5} + 16$;

thus for a wheel 15ft. diameter, the number of floats
$= \dfrac{8 \times 15}{5} + 16 = 40.$

Breast Wheels.

The essential difference in the actions of water on undershot wheels, and on overshot and breast wheels, is that in the undershot form the wheel is driven by the impulse of the water on the float-boards, and in the other two forms it is turned by the weight of the water. For equal quantities of water expended the action by weight is about twice as effective as by impulse. An ordinary "undershot" will raise back again to the height from which it falls about one-third of the quantity of water expended; Poncelet's wheel fully one-half; while an ordinary "overshot" will raise three-fifths; an overshot wheel of good construction will raise from two-thirds to seven-tenths; and a high-breast wheel of the best construction will raise three-fourths of the quantity of water expended, back again to the height from which it falls. A high-breast wheel is formed with buckets, similar to those of an overshot, which retain the water poured into them above the axis of the wheel, acting upon it by the force of gravity. In a low-breast wheel the water is applied below the level of the axis, and is made also to act by its weight by making the wheel to move in a close-fitting race of masonry, which holds up the water to the breast of the wheel.

In order that the water may act upon the wheel by weight through the greatest fall it is made to run over the top of the gate in the form of a weir, the gate being lowered sufficiently to allow the proper quantity to flow over it, and raised to shut off the water; called the sliding hatch. To ascertain the velocity and quantity of water flowing over such a sluice-gate in a given time, reference must be made to experiments on the flow of water over weirs. The flow over weirs follows the same general law as that which governs the flow through openings below the surface; that is to say, the velocity is proportional to the square root of the head of water, or the depth in the case of weirs. The theoretical velocity in both complete orifices and weirs is $8\cdot 03 \sqrt{h}$, h being the height in feet from which the water falls, or the head of water (reduced from the theorem of Torricelli, $v = \sqrt{2gh}$) in which $v =$ the velocity in feet per second, and $g =$ the force of gravity, or the velocity acquired by a heavy body at the end of the first second of its time of falling, $= 32\cdot 2$ft. per second; but the practical velocity is less; under the most favourable circumstances it is $7\cdot 8 \sqrt{h}$, but in general the value of the co-efficient may be taken to be $7\cdot 5$.

The quantity discharged is diminished in practice in both kinds of openings by the same influences; viz., by the contraction of the stream caused by projections into it, which diminish its sectional area, and of the friction on the border of the stream, which diminishes the mean velocity.

The speed of the wheel should be somewhat less than the velocity of the water in order that the float-boards may not strike against it and retard the motion of the wheel. As to the depth or head of water, it is to be remarked that it is not the depth measured immediately over the gate or hatch itself, but from the level of still water a few feet away from it; and the depth from which the velocity is to be calculated is the whole depth from the surface of still water to the top of the gate. The actual depth measured on the gate is, on an average of all depths, about four-fifths of the full depth, approximately.

The mean velocity of the stream going over the gate, calculated from the depth between the still water of the head and the top of the gate, is two-thirds the velocity due to the lowermost thread of particles, which is $7\cdot 5 \sqrt{h}$. That thread of particles which has the mean velocity of the whole is at a depth below the surface of four-ninths of the whole depth, and is

$$7\cdot 5 \sqrt{\frac{4}{9} h} = 7\cdot 5 \times \frac{2}{3} \sqrt{h} = 5 \sqrt{h}.$$

If, for instance, the depth were 6in., or ·5ft., the mean velocity would be $5 \sqrt{·5} = 3·54$ ft. per second. But the actual velocity of the sheet of water passing over the gate would be greater in the ratio of 5 to 4, for the depth of water on the gate would be approximately four-fifths of the whole depth. Thus the actual velocity would be $\dfrac{3·54 \times 5}{4} = 4·42$ft. per second; and the speed of the wheel might be 4ft. per second, so that the spaces between the float-boards which are successively presented to the flow of water may be fully supplied. The speed, however, may be required to be greater than 4ft. per second, according to the kind of work it has to do, and according to the intermediate gearing.

Sir William Fairbairn usually made the speed of breast wheels from 4 to 6ft. per second at the periphery. If the water does not flow immediately on to the wheel from a still head, but arrives in a channel of no great width, and therefore with a perceptible velocity, let that velocity be ascertained with floats and reduced to feet per second; or if not by direct observation with floats, let the known approximate quantity of water coming down the channel be divided by its sectional area, which will give the mean velocity of the stream, and that velocity increased by one-fourth will in most cases give the velocity at the surface, which is that to be ascertained, and if it be called w, then v, the mean velocity due to the whole depth of water above the gate, will be $5 \sqrt{h + ·035 w^2}$, and that of the sheet of water passing over the gate will be greater than this in the ratio 5 : 4.

The parts of a low-breast wheel are, besides the axle and the arms, the rim, the starts, the sole-boards, and the float-boards. The sole-boards should not close up the spaces between the float-boards, but at the top of each a horizontal slit of an inch or so in width should be left for the escape of air when the water flows in, and for its admission when the water is discharged after passing the vertical centre of the wheel, otherwise the water will neither enter freely nor be discharged freely. The axle, arms, and starts, are usually of oak or elm; the float-boards of fir.

Another form of sluice gate is the double hatch, the opening being wholly below the surface, and the upper and lower parts of the gate being brought within the required distance of each other by racks and pinions. This form is adopted to ensure a greater velocity in the issuing stream of water than it can attain when it passes over the top of the gate, with the same expenditure of water. The quantity of water discharged in a given time by either form of gate may be calculated on the same basis. It is proportional to the square root of the height from which it falls and to the area of the opening through which it issues—to the velocity and sectional area combined. Water falling over a weir takes the form of a parabolic curve, and there issues in any given time two-thirds of the quantity which would issue in the same time from an opening of the same dimensions placed horizontally at a depth below the surface equal to the whole depth of water flowing over the weir. Through an opening wholly below the surface the quantity is proportional to the area multiplied into the velocity. The causes of diminution of the flow in either case are (1) contraction of the stream, and (2) friction, or that retardation of the particles of water next the bottom and sides which ensues upon their being rolled over each other, by reason of the contact of some of them with the confines of the channel, while those in the axis of the stream pursue a straighter course.

CURRENT WHEELS.

The original form in which use was made of water in motion to do mechanical work was that of opposing to the force of the current of a river a series of float-boards placed round a wheel. In Ewbank's historical account of machines for raising water he describes the tympanum (Fig. 37) used by the Romans, and so called by them from

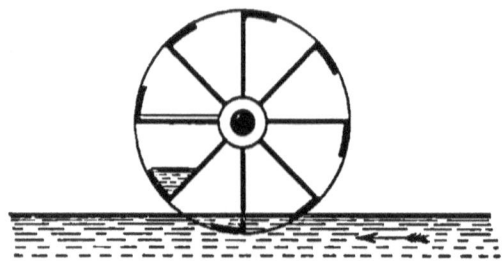

Fig. 37.

its resemblance to a drum; but a similar form of wheel had been used in more ancient times. It had a hollow shaft, or axle, and the arms proceeding from it to the circumference of the drum were bent up at their ends, and formed so many scoops, which, when the wheel was turned round, gathered up each its portion of water, and as each arm came to a horizontal position the water was shot inwards and entered the hollow shaft, and, one end only being open, the water was discharged from it and led away.

This machine was greatly improved in the early part of last century by M. De La Faye, a member of the French Academy of Sciences, who "developed by geometrical reasoning a beautiful and truly philosophical improvement." Quoting from Belidor's description of De La Faye's reasoning, Ewbank gives it thus:—"When the circumference of a circle is developed a curve is described (the involute), of which all the radii are so many tangents to the circle, and are likewise all respectively perpendicular to the several points of the curve described, which has for its greatest radius a line equal to the periphery of the circle developed. Hence, having an axle whose circumference a little exceeds the height which the water is proposed to be elevated, let the circumference of the axle be evolved, and make a curved canal, whose curvature shall coincide throughout exactly with that of the involute just formed. If the further extremity of this canal be made to enter the water that is to be elevated, and the other extremity abut upon the shaft which is turned, then in the course of rotation the water will rise in a *vertical direction*, tangential to the shaft and perpendicular to the canal, in whatsoever position it may be." As given by Belidor (Fig. 38) it has four arms, but Ewbank says that is frequently constructed with double that number.

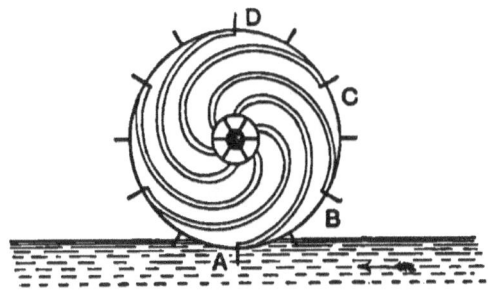

Fig. 38.

Dr. Olinthus Gregory, in his treatise on Mechanics, after describing this same machine, says:—"Thus the action of the weight, answering always to the extremity of a horizontal radius, will be as though it acted upon the invariable arm of a lever, and the power which raises the weight will be always the same; and if the radius of the wheel, of which this hollow channel serves as a bent spoke, is equal to the height that the water is to be raised, and consequently equal to the circumference of the axle or shaft, the power will be to the load of water reciprocally as the radius of a circle to its circumference, or directly as 1 to $6\frac{1}{4}$ nearly." The wheel being turned by the impulse of the stream acting upon the float-boards, the orifices, A, B, C, D, &c., of the curved arms dip, one after another, in the water, which runs into the channels so formed, and as the wheel revolves the fluid rises in them vertically and enters the hollow shaft, from which it runs out in a stream at either end.

This form of wheel can only raise water to the height

of half its diameter, but the "Persian Wheel" (Fig. 39) takes it up nearly to the full height of its diameter. Buckets are hung upon pins projecting from one or both sides of the wheel, and when they arrive at the top they are emptied by coming in contact with the trough, which conveys the water away. For this purpose they are furnished with bowed springs, which come in contact with a fender at the side of the trough into which the water is discharged, and so the bucket is tilted up and emptied, and swings down again to its vertical position.

The power of such wheels as these is measured by the area of the float-board and the velocity of the current combined. The velocity is most easily ascertained by

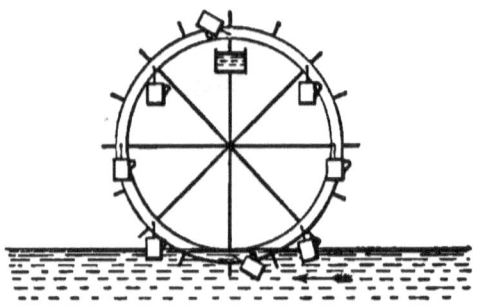

Fig. 39.

means of floats, which are sometimes of wood, painted white and loaded with lead until the upper surface stands nearly level with the surface of smooth water; but, perhaps, spheres are better, of some material which displaces nearly, but not quite, its own bulk of water, as offering less obstruction to the action of wind. Oranges are sometimes used.

Having ascertained the surface velocity of the stream, the inquiry would be—What produces this velocity? It is the force of gravity acting on a mass of matter which possesses the quality of fluidity; and water is drawn to the ocean from its source by that power, and is interrupted in its course thither by being employed to turn wheels, or otherwise to transmit the force inherent in it to the work to be done. The surface of a stream of water may be considered as an inclined plane, reversed, down which moves the body of water. When water enters a river or stream, time must elapse before that portion of the river or stream below the point at which the water enters becomes *in train*. A river is in train when the inclination of its surface does not vary while its hydraulic mean depth remains constant; and it may be said to be in progressive train and in retrogressive train when the inclination and hydraulic mean depth increase gradually and decrease gradually, respectively. For instance, after the flood waters of the upper part of a river had ceased to flow into it, it would not become in train until all that water had spread its influence over the lower part of the river; but after that it would continue in train until it should be again influenced by additional water. Whenever that might occur, if the quantity were to run in gradually until it were exhausted, the river would be put into a state of progressive train, and if, by reason of no more water running into the river until that which had run in had run out, the river would be in the meantime in a state of retrogressive train; and so the regimen of the river might fluctuate gradually in this way; but it is seldom that this regular influx and efflux take place.

The regimen is usually interrupted by intermediate additions of water, so that, except in dry weather, it is not easy to ascertain when a river is or is not in train, and the alternative is to observe to what level the surface marks in summer, excluding excessive droughts, and to call that the summer level. The river is at those times "in train." Taking the surface of the river, then, when it is in train, the water drawn along it by the force of gravity may be said to move along the inclined path of its surface reversely, for the same reason that a solid moves down and upon an inclined path; and that, of the laws governing the two cases, some are applicable to both, while others are applicable to each respectively.

If an inclined path down which a solid body might

move were perfectly smooth, there would be no sliding friction, and as soon as the limiting angle of resistance to the friction of stability should be attained, the solid would shoot down the incline with an accelerating velocity. In a river the water is not perfectly fluid, but has some viscidity. If it were a perfect fluid it could not be retained in its channel, and no state of the river such as that we call in train could be maintained.

It is sometimes said that it is "fortunate" that water decreasing in temperature below about 40 degrees Fahr., expands, and that when it is converted into ice the specific gravity is so much less than that of water that it floats upon it, otherwise that the whole would become a solid mass of ice. It may be said to be equally "fortunate" that water is not a perfect fluid, otherwise it would all run out of the river in a glut, and no permanent stream could be maintained.

When water is gathered together in a stream and descends towards the sea, it does so with a velocity the square of which is proportionate to the inclination of its surface, combined with its hydraulic mean depth, the mean velocity of the whole body being in general 8 or 9 tenths (according to the roughness of the bed) of a mean proportional between its hydraulic mean depth and twice the fall of its surface per mile. Water flowing in the axis of a stream goes down with greater velocity than that near the sides, which shows that, although we say water is a perfect fluid for all practical purposes, it yet possesses some viscidity, for its outer portions hang to the ground over which it runs, and the velocity increases from the bottom upwards and from the sides inwards, until the maximum velocity is attained in the axis of the river; and it is even observed in parts of some rivers that there is a reverse current near the shore, running up the river. Between this upward current and the main downward current there are eddies. There are also eddies at some distance upwards from the point of observation of the shore water, at which point it turns downwards again and gets into the main stream. The velocity is greatest in the axis because the channel is there the deepest.

The axis need not be in the centre of the river, for it is its main channel, wherever that may be; and the velocity is there the greatest, because at any point of the depth— whether at the point where the mean velocity takes place or whether at the surface—it is further removed from the retarding influence of the bed. The motion is due to the attraction of the mass of the earth, and it finds the readiest way to the lowest point continually. The velocity it acquires in falling is the measure of its force. It rolls with an involved motion, the moving particles at the surface being continually replaced by those from below and from the sides. All these things show that water is not a perfect fluid, but that it has in some degree the quality of viscidity, or viscosity, if that word more exactly explains the meaning.

The wheels last described are current wheels, and there are many situations where they may be usefully employed, especially in new countries, and where other forms could not be readily adopted. The power of such wheels may be determined on the following considerations:—If the wheel be fixed, and the current of water impinge against the float-boards it will do no work, but the effort of the water, tending to turn the wheel round, will be the greatest possible—will be the maximum effort that the water is capable of when running with its actual velocity. On the other hand, if the wheel be freed, and it offer so little resistance to the current of water as to move with it —that is, with the same velocity—then again no work will be done, there being no resistance.

Between these two states there is one which produces the maximum effect, or greatest amount of work. When the water is at liberty to escape sideways as fast as it impinges on the float-boards, as it does in current wheels, it can be shown by a process of reasoning that the effect is a maximum when the float-boards move with a velocity equal to one-third of the velocity with which they would move if there were no resistance, or one-third of that of the current of water; and also that the most effective force with which the water can act on the wheel is four-ninths of its utmost force when its head accumulates against the float-board when the wheel is held still. This force is equal to the weight of a column of water whose base is the area of the immersed float-board, and whose height is that due to the velocity of the current. That height is found thus. The force of gravity constantly adds a certain quantity of motion, second by second, to a body falling freely, whatever actual velocity it may have attained at the end of any previous second, and this quantity is $32\frac{1}{6}$ft., and is usually represented by the letter g. In the first second, after falling from a state of rest, it falls a distance of $16\frac{1}{12}$ft., being the mean between zero when at rest and $32\frac{1}{6}$ft., the velocity it has acquired at the end of the first second. For practical purposes such as these we are considering, it is sufficiently accurate to take $g=32$, and h, the height fallen through in the first second, $=16$. The velocity will then be for any height, $v = \sqrt{2gh}, = \sqrt{64h} = 8\sqrt{h}$; and conversely, the height corresponding to any given velocity, v, will be $h = \dfrac{v^2}{64}$, v being the velocity in feet per second. Supposing, then, the velocity of any stream employed to turn a current wheel to be found by experiment to be 8ft. per second, the height from which it must have fallen to produce this velocity is $\dfrac{8 \times 8}{64} = 1$ft. Supposing the immersed area of the float-board to be 10 square feet, the force of this current on this immersed area of float-board would be $10 \times 1 \times 62\frac{1}{2} = 625$lb. But this is the whole force due to the stream of water, and can only take effect when the wheel is held still and therefore, of course, is doing no work. When let go it will move with a velocity, compared with that of the stream, which will be due to the extent to which it may be loaded, and the maximum effect, or greatest amount of work done, will take place when the force acting on the float-board is $\frac{4}{9}$ of the whole force of the stream, or $\left(\frac{4 \times 625}{9} =\right)$ 277·7lb., and when the velocity of the wheel is $\frac{1}{3}$ of that of the water, or $\left(\frac{8}{3} =\right)$ $2\frac{2}{3}$ft. per second, the work done will be that of falling (277·7 × $2\frac{2}{3}$ =) 740lb. 1ft. per second, which is about $1\frac{1}{3}$ horse-power. This takes account of only one float-board, but it has been found by experiment that by making the number of float-boards greater than is strictly required in order to keep one float-board always fully immersed, a greater effect is produced, and taking this practical increase of effect into account we may say this wheel would be, at least, $1\frac{1}{2}$ horse-power.

Where a stream can be confined in its channel so as to make all the water that impinges on the float-boards continue to act upon them until the wheel releases it, the best relative velocity of the wheel to that of the stream is one-half, instead of one-third when unconfined. But when the circumstances of the situation are such that an artificial channel can be made to bring the water to the site of the work the wheel is taken out of the category of current wheels and becomes a proper "undershot" wheel; as, for instance, where a stream, being small, may be wholly dammed up for the purpose of gaining a few feet head of water without causing it to overflow the banks of the stream above, or do other damage; or where part of the water of a larger river may be diverted from a point at some distance above the site of the intended wheel and brought to it in an artificial channel having a less rate of inclination than the river itself, so that a few feet head of water can be accumulated, then a proper undershot wheel may be applied.

SECTION XIX.

CORN-MILLS.

THE cases to be presently mentioned are such as admit of the quantity of water expended being ascertained, which is usually the whole volume of the stream, and whether that be so or not in particular cases the quantity expended can be ascertained from the two measurements of the fall and the opening of the sluice-gate; but current wheels such as those last mentioned are sometimes applicable.

The quantity of water which actually strikes the wheel is always less than the whole quantity which comes against it; it bears the same relation to the whole quantity as the difference between the velocity of the water and that of the wheel bears to the absolute velocity of the water. The force with which this quantity strikes the wheel is as the square of the relative velocity, or the difference between the two velocities, for the force is as the relative velocity multiplied into the quantity striking the wheel with that velocity, and the quantity is as the relative velocity, therefore the force is as the square of the relative velocity. If, for instance, the stream move at the rate of 9ft. per second, and the wheel at the rate of 3ft. per second, 36, the square of the difference, represents the force acting upon the wheel; but if in the same stream the wheel move at the rate of 6ft. per second, the force is only 9, the square of 3, the difference between the two velocities. But when a close race is fitted to the lower portion of the wheel, which is the usual mode of construction, so that the water does not all escape immediately after striking the float-board with which it comes in contact, but continues to act until it finally leaves the wheel at or near the bottom, the case is different, and under these circumstances the best velocity for the wheel is found to be nearer one-half than one-third of that of the water.

In general, the largest bodies of water are found in the lowest positions, with small or but moderate falls, and although the application of water to turn the wheels is better effected with a steady pressure and uniform motion —that is by its weight—yet when the quantity is large, and the fall but little, it cannot be so well applied in that way as by means of undershot wheels. The water acts upon these by impact, in contradistinction to weight, and the power of water acting in this manner is derived from a force different in its nature from that which produces a steady pressure.

Taking the quantity of water expended to be the same in the two cases, it would be discharged in the former through a small opening with great velocity, and in the latter case through a large opening with but little velocity; in the one case it is discharged near the bottom of the fall, in the other near the top. At a depth of 5·76ft. the velocity through a sluice is $7·5 \sqrt{5·76} = 18$ft. per second. The constant 7·5 is thus derived. In the discharge of water through holes in thin plates the stream is found to be contracted in sectional area at a short distance outside the opening. With round holes the diameter at the contracted part can be measured, and its sectional area ascertained, and it has been found by hydraulicians who have made these measurements that the area at the contracted part of the stream is about two-thirds of that of the hole. If the velocity at this part suffered no diminution, which is sometimes assumed, the discharge would be $a \times 8 \sqrt{h}$, a being the sectional area of the stream at its smallest part, and h the height of the surface of the water above the centre of the opening; and if A be the area of the opening, the quantity issuing would be $\frac{2}{3}$ A × 8 $\sqrt{h}$. But the cubic quantities actually measured do not amount to so much as this represents; they amount to only $\frac{5}{8}$ A × 8 $\sqrt{h}$; the difference, therefore, must be caused by a diminution of the velocity in the ratio $\frac{2}{3}$ to $\frac{5}{8}$, or 8 to 7·5, in the contracted part of the stream; and in the opening itself in the ratio 1 to $\frac{5}{8}$ or 8 to 5; and the diminution of velocity in the contracted part is probably the same in all forms of opening, although the degree of contraction varies much with the forms; thus the actual velocity of a stream issuing from under a sluice-gate, taken at that part where it is greatest, is 7·5 $\sqrt{h}$, h being measured in feet, and the velocity in feet per second. The most general proportions of velocities between the water and the wheel found in twenty-seven experiments by Smeaton was 10 to 4, the extremes being 10 to 3·4 and 10 to 5·2; and the most general proportion between the power and effect was 10 to 3, the extremes being 10 to 3·25 and 10 to 2·82; but as the former is that which obtained when the power exerted was greatest, the proportion in large works may properly be taken at 3 to 1 for ordinary undershot wheels with straight float-boards, where the water is confined to them by a close race.

Amongst the trials I have made of the quantity of water expended in doing various kinds of work may be selected, first, either those in respect of different kinds of work on the same kind of wheel, or those in respect of the same kind of work on different kinds of wheel: the latter method seems preferable, and the following are selected from some trials of the quantity of water expended in grinding corn with undershot wheels. These, as will be seen, are much less effective, in proportion to the quantity of water expended, than either overshot or breast wheels; but there are many of them, and notwithstanding their less effectiveness, they are more appropriate for large quantities of water and low falls than the other kinds of wheel.

In a corn-mill driven by an undershot wheel 16·80ft.

diameter, 8·50ft. wide, which has 24 straight float-boards 2·25ft. wide, without soling, there are two pairs of French burr wheat-stones, 4ft. 10in. diameter, and one pair 4ft. 8in. diameter; also one pair of grey stones 5ft. diameter, one pair of shelling-stones, one corn-screen, and two flour-dressing machines. The centre of the wheel is 2·20ft. above the level of a full head of water. The bottom of the sluice-opening is 3·40ft. below full head. When the wheel is standing, the tail-water is 6·20ft. below full head. On the occasion of the following trial, No. 1, the water was ·56ft. below full head, the height above the bottom of the sluice-opening being 2·84ft. Two pairs of 4ft. 10in., and one pair of 4ft. 8in. wheat-stones, and one flour-dressing machine, are running. The sluice-gate is drawn ·81ft. The surface of the water in the tail-race is 5·20ft. below the full head, or 4·64ft. below the present head. The head above the centre of the sluice-opening is $2·84 - \frac{·81}{2} = 2·44$ft. So far, these are matters of observation and measurement, but the next is one of calculation;—that is, the quantity of water discharged through the sluice-opening, which is 8·70ft. wide, and formed as in the diagram. There are side walls, as in Fig. 32, and the bottom is level with the sill. The side walls are not of the best form for facilitating the passage of the water, but they conduce to it in some degree. Under the circumstances, the proper co-efficient of discharge would probably be ·75, and the quantity of water $8·70 \times ·81 \times 8 \sqrt{2·44} \times ·75$ = 65 cubic feet per second, or 3,900 cubic feet per minute, which, multiplied into the fall, 4·64ft. = 18,096. There being three pairs of stones running, the quantity of water per minute per foot of fall per pair of stones is 6,032 cubic feet, including the dressing machinery.

The next trial, No. 2, was with a full head of water, with the same wheel, the machinery running being two pairs 4ft. 8in. and one pair 4ft. 10in. wheat-stones, one pair 5ft. grey stones, and one flour-dressing machine. The sluice-gate was drawn ·76ft. There being a full head of water the surface water was $3 \cdot 40 - \dfrac{\cdot 76}{2} = 3 \cdot 02$ ft. above the centre of the opening. The area of the opening is $8 \cdot 70 \times \cdot 76 = 6 \cdot 60$ sq. ft., and the discharge would be $8 \cdot 70 \times \cdot 76 \times 8 \sqrt{3 \cdot 02} \times \cdot 75 = 68$ cubic feet per second, or 4,080 cubic feet per minute. There being a full head, the fall was $5 \cdot 20$ ft., and $4,080 \times 5 \cdot 20 = 21,216$. If the driving of the meal-stones requires as much power as a pair of wheat-stones, so that the number might be reckoned as four pairs, then the quantity of water expended per minute per foot of fall per pair of stones is 5,304 cubic feet; but if the number be taken at $3 \cdot 5$ pairs, then 6,060 cubic feet. On the third trial with this wheel the water was $\cdot 14$ ft. below full head, and the sluice-gate was drawn $\cdot 80$ ft., the machinery running being the same as on the last trial. The height of the water above the bottom of the sluice was $3 \cdot 40 - \cdot 14 = 3 \cdot 26$ ft. The height above the centre of the opening $3 \cdot 26 - \cdot 40 = 2 \cdot 86$ ft., and the discharge $8 \cdot 70 \times \cdot 80 \times 8 \sqrt{2 \cdot 86} \times \cdot 75 = 70$ cubic feet per second, or 4,300 cubic feet per minute. The fall was 5ft., and $4,300 \times 5 = 21,500$. If the work done be reckoned as four pairs of stones, the quantity per pair is 5,375 cubic feet; but if as only $3\frac{1}{2}$ pairs, then 6,143 cubic feet per minute per foot of fall, the dressing machinery being included. The speed of this wheel is ten revolutions per minute. On the fourth trial the grey stones and the dressing machine were thrown off, and the three pairs of wheat-stones run. The water was $1 \cdot 50$ ft. below full-head, and the sluice-gate was drawn $\cdot 98$ ft. The height of the water above the bottom of the sluice was $3 \cdot 40 - 1 \cdot 50 = 1 \cdot 90$, and above the centre of the opening $1 \cdot 90 - \cdot 49 = 1 \cdot 41$. The discharge $8 \cdot 70 \times \cdot 98 \times 8 \sqrt{1 \cdot 41} \times \cdot 75 = 61$ cubic feet per second, or 3,660 cubic feet per minute. The fall was $5 \cdot 20 - 1 \cdot 50 = 3 \cdot 70$ ft., and $3,660 \times 3 \cdot 70 = 13,542$.

There being three pairs of stones running, the quantity of water expended per pair of stones would be 4,514 cubic feet per minute per foot of fall.

TRIALS OF QUANTITY OF WATER USED.

At another corn-mill there are three wheels, the trials with one of which only are selected. It is 16·30ft. diameter, 6·20ft. wide, with 24 straight float-boards 2·25ft. deep. The centre of the wheel is 0·56ft. above the level of a full head of water. The bottom of the sluice-opening is 5·61ft. below full head.

On the occasion of the following trial, No. 1, the water was 1·15ft. below the level of full head. The sluice was drawn ·98ft. The head above the bottom was, therefore, $5 \cdot 61 - 1 \cdot 15 = 4 \cdot 46$ft., and above the centre of the opening $4 \cdot 46 - \frac{\cdot 98}{2} = 3 \cdot 97$ft. The width of the sluice-gate is 6·70ft. The discharge, therefore, would be $6 \cdot 70 \times \cdot 98 \times 8 \sqrt{3 \cdot 97} \times \cdot 75 = 76$ cubic feet per second, or 4,560 cubic feet per minute. The fall from present head to tail-water is 5·46ft., and $4,560 \times 5 \cdot 46 = 24,897$. There were four pairs of stones running at the time, and the quantity per foot of fall per minute per pair of stones would be 6,224 cubic feet. The speed of this wheel is eight revolutions per minute.

On the next trial, No. 2, with the same wheel, the water was 1·35ft. below full head, the sluice was drawn 1·26ft., the head above the bottom was $5 \cdot 61 - 1 \cdot 35 = 4 \cdot 26$, and above the centre of the opening $4 \cdot 26 - \frac{1 \cdot 26}{2} = 3 \cdot 63$ft. The discharge would be $6 \cdot 70 \times 1 \cdot 26 \times 8 \sqrt{3 \cdot 63} \times \cdot 75 = 95$ cubic feet per second, or 5,700 cubic feet per minute. The fall was 5ft., and $5,700 \times 5 = 28,500$. There being four pairs of stones, the quantity is at the rate of 7,125 cubic feet per pair.

The next trial was with a wheel where the sluice-opening is 10·50ft. wide. The depth of water to the bottom of the sluice was 5·45ft. The gate was drawn ·68ft. The head upon the centre of the opening was, therefore, $5 \cdot 45 - \frac{\cdot 68}{2} = 5 \cdot 11$ft., and the quantity discharged $10 \cdot 50 \times \cdot 68 \times 8 \sqrt{5 \cdot 11} \times \cdot 75 = 96$ cubic feet per second, or 5,760 cubic feet per minute. The fall was 5·32ft. at the time of the trial, and $5,760 \times 5 \cdot 32 = 30,643$. The number of pairs of stones being five, the quantity per pair was 6,128 cubic feet.

All these wheels are undershot, and of rather rude construction, but not, perhaps, unusually so.

The next trial was with two low-breast wheels 14ft. diameter, each 10ft. 6in. wide, where the water falls over the top of the gates. The combined length of the opening is 19ft., and the depth of water going over $6\frac{7}{8}$in.

The form of the tops of the gates is such as to make it probable that five would be very nearly the proper constant to apply for the quantity of water in cubic feet per minute, and $5 \sqrt{d^3} \times l = 5 \sqrt{(6\frac{7}{8})^3} \times 19 = 1,700$.

$1,700 \times 5$ft. fall $= 8,500$. There were running three pairs of wheat-stones 4ft. diameter, with the dressing machinery, which would make, per pair of stones, 2,833 cubic feet per minute per foot of fall.

The useful effect of the weight of a given quantity of water acting during a long time, the velocity being consequently small, is much greater than that of the same quantity acting by the impulse of a great velocity. If the quantity per second or per minute be multiplied into the vertical height of the fall, and called M, this divided by 528 exhibits the horse-power of the water when applied by its weight upon the wheel; but when it acts by impulse, M must be divided by 1056 for the horse-power of the water, and if it acts through part of the fall by impulse, and through the remainder by weight, these parts must be taken separately. The horse-power of a waterfall, therefore, cannot be stated, even when the quantity and fall are known, without taking into account its mode of action, for it requires twice as much water to produce a given power by way of impulse, as that which is required to produce the same power by its weight, moving with slower speed. If, for instance, Q be the quantity of water in cubic feet per minute, and h the height of fall in feet, the horse-power is $\dfrac{Qh}{528}$ when its weight is brought to bear upon a wheel which has a velocity of its circumference sufficient only to give the water room to fall freely; and the slower the speed the greater the useful effect, according to the following reasoning. (Robison's 'Mechanical Philosophy.')

Putting the work to be done into the form of a weight to be raised, "if R be the radius of the wheel to the centre of gravity of the water in the bucket, and w be the weight of the column of water acting at that distance from the centre of the wheel, and if r be the radius of the axle, or the distance from the centre at which the weight W reacts upon the falling water, then the forces are in a state of equilibrium when $R\,w = r\,W$, and this is so whether the acting and reacting forces be at rest or move with any velocity whatever, so that it be uniform; for gravity would accelerate the falling water if it were not completely balanced by some reaction, and in this balance, gravity and the weight raised exert equal and opposite pressures, and thus produce uniform motion. Now as to the speed, both the falling water and the weight raised may be taken for comparison in cubic feet of water, and if A be the cross-sectional area of the falling column of water, and h its height, its weight may be represented by $A\,h$.

"If V be the velocity of the descending weight w, in feet per second, and v that of the ascending weight W, then $Wr = AhR$; but $R : r :: V : v$, therefore $Wv = AhV$. The work done is measured by the weight and the height to which it is raised per second, or by Wv; therefore, the greatest amount of work is done when AhV is a maximum. But AV is a constant quantity, being the quantity of water descending per second; if the wheel moves fast A is small; when V is small A is great, but AV remains the same; therefore, h should be the greatest possible—that is, the water should be applied upon the wheel as high up as possible; but this diminishes the head necessary to give the water its velocity of entry into the buckets, which must be such as to prevent them striking it in passing the entering stream of water; and a diminution of this head diminishes the velocity of the entering water, while, at the same time, the less the speed of the wheel the less need this velocity be. As the diminution has no limit, the reasoning shows that an overshot wheel does more work as it moves with slower speed." But practical considerations and experiments have shown that a velocity of at least 3ft. per second should be given to the circumference of a water-wheel, and that it may be as much as 6ft. per second without much loss of effect.

With overshot wheels it is very desirable to keep the water up to the same head at all times; but this cannot

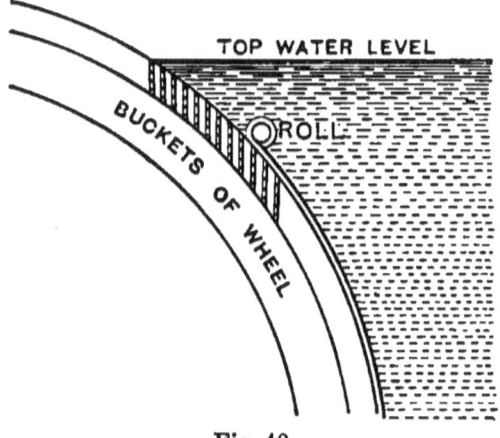

Fig. 40.

be done without wide reservoirs. The small reservoirs or dams for the storage of the night water, or at most, for the 36 or 40 hours from Saturday to Monday, are not of sufficient capacity to do this, and the head consequently varies considerably between morning and night, as well as at different times of the year. There are two ways of applying the water under these circumstances; one by means of a sliding gate or roll of leather over which the water flows upon the wheel at various heights, but always with the same velocity, and therefore without unnecessary loss of head (Fig. 40), and the other by means of a pentrough above the wheel to contain a considerable depth of water (Fig. 41), whereby a sacrifice of power is made when the water is at a high level, the diameter of the wheel and the application of the water being accommodated to the lowest working head. The following case is one of this sort:—The former method is much the better, but in water questions it is not always what is best that can be insisted upon, but what exists which must be dealt with. A corn-mill has an overshot wheel 15ft. diameter, 7½ft. wide. The wheel works close under

Fig. 41.

the pentrough, as in Fig. 40, and the water is, therefore, laid on at the highest possible point, which is about 25 deg. from the vertex. At the time the following observations were made there was a depth of water in the pentrough of 4ft., the sluice-opening being in the bottom, which consists of 3in. planking. The length of the opening is 6·40ft., and the width, when all the machinery of the mill is running, is 3in.; the area of the opening is, therefore, 1·60 sq. ft. The coefficient of discharge through such an opening would be about ·60, and the quantity of water passing through the sluice would be $\sqrt{64 \times 4} \times 1\cdot 60 \times \cdot 60 = 15\cdot 36$ cubic feet per second, or 920 cubic feet per minute. The water is laid on to the wheel 1ft. below the bottom of the pentrough, and about 1½ft. is lost at the bottom of the wheel, the effective fall being $15 - 2\cdot 5 + \dfrac{5}{2} = 15$ft. The power of the water applied through this wheel is, therefore, that of 920 cubic feet per minute falling 15ft., or 13,800 cubic feet per minute falling 1ft.

The machinery in the mill, all of which can be run at the same time with the before-named quantity of water, is: Four pairs of French burr wheat-stones, 4ft. 4in. diameter; one pair of Derbyshire grit meal-stones; one pair of shelling-stones; one bean-splitter; one corn-dressing machine; one flour-dressing machine, 20in. diameter, 6ft. long. If the other pairs of stones be supposed to require the same power as the wheat-stones, and that there are in all six pairs, $\dfrac{13,800}{6} = 2,300$ cubic feet of water per minute per foot of fall for each pair; and $\dfrac{2,300}{528} = 4\cdot 35$ horse-power.

The quantity of wheat ground per hour is about four bushels by each pair of stones, when in good condition; or 3½ bushels at other times.

The second example is that of a flour-mill. There is a slight difference, it may be said parenthetically, between a flour-mill and a corn-mill. In country places, such as that where the mill first mentioned is situated, other grain besides wheat is ground for the convenience of the neighbourhood, and for these the term corn-mill is more appropriate; but mills near large towns are often fully occupied in grinding wheat, except that they may have a pair of meal-stones for occasional use, and these are more properly called flour-mills. In the following case there are two breast-wheels, each 16ft. diameter, one 12ft. wide, the other 8ft., with cast-iron hollow shafts, and cast-iron arms and rings, into which oak starts are fitted. The larger, or No. 1 wheel, has four sets of arms and rings, and forty oak starts in each ring; the smaller, or No. 2 wheel, has three sets of arms and rings, and the same number of starts in each ring as the other wheel has; and there are 40 float-boards of elm in each wheel, 18in. in depth, and 40 back-boards, also of elm. The circumference of the wheels being 50ft., the floats are 15in. apart from centre to centre. When the following observations were made, the total fall of water was 6ft. between the head and the surface of the water in the tail-race when the mill was running, and as the tail-race is carried up to the centre of the wheel or nearly so at a good depth, about 2ft., there is no drag on the wheels by running in backwater. The wheels have cast-iron breasts, which confine the water closely to the edges of the float-boards, and curved water-gates of elm, worked by racks and pinions. The water-gate of the large wheel is double, the combined width of the two openings being $11 \cdot 30$ft.; the width of that of the smaller wheel is $7 \cdot 90$ft., and the water runs over the top of the gate, and under a fixed head-beam, as shown in Fig. 42.

There are ten pairs of stones in this mill, nine pairs of French burr wheat-stones, 4ft. 2in. diameter, and one pair of Derbyshire grit meal-stones, with all the necessary wheat-dressing and flour-dressing machinery. But at the time of the experiments there were only seven pairs of wheat-stones running, together with the dressing-machinery, or as much of it as was required to keep in work the seven pairs of stones. The quantity of water is calculated in the following manner, which, for small heads above the openings, is more accurate than that of measuring the head of water from the surface to the centre of the opening. The gate being lowered by the miller to the depth which he finds sufficient to do the work, the heights H and h in Fig. 42 are measured, after a sufficient time has elapsed to allow the water to settle to a steady head, and after the miller has finally adjusted the height of the gate. As the thickness of the water-gate is 6in., and the side and division-posts are square in form, the proper co-efficient of discharge would be about ·63. The quantity of water passing through such an opening is taken to be the difference between those which would pass over two weirs, the one having the depth H, and the other the depth h, both below the surface of still water. In this way, on the first experiment of No. 1 wheel, the depth H was 11¼in., and h 4in., the combined length of the two openings being 11·30ft.

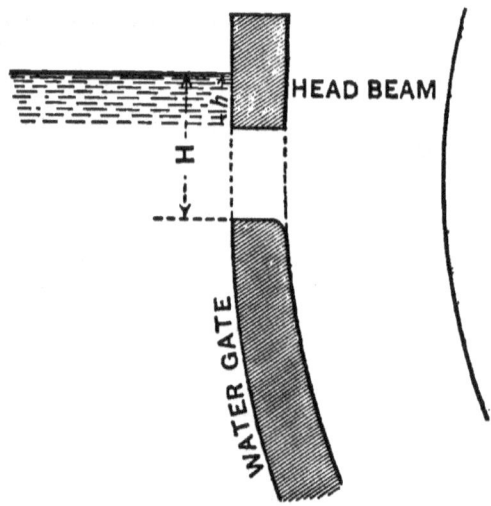

Fig. 42.

The fundamental condition is that at the depth H or h, the velocity of the water is proportional to $\sqrt{64\,H}$ or $\sqrt{64\,h}$, and the mean velocity of the whole sheet of water is two-thirds of that which it is at the depth H or h, the quantity actually discharged being corrected by the co-efficient ·63, which embraces the corrections for both retardation of velocity and contraction of the sectional area of the stream, caused by its passing through the square form of opening, as in this case. Generally, if l be the length of opening in feet, the quantity discharged in cubic feet per second would be

$l \text{H} \times \frac{2}{3} \sqrt{64 \text{H}} \times \cdot 63 - lh \times \frac{2}{3} \sqrt{64 h} \times \cdot 63$

which is more conveniently stated as

$5 \cdot 33 l \sqrt{\text{H}^3} \times \cdot 63 - 5 \cdot 33 l \sqrt{h^3} \times \cdot 63;$

or, finally, as

$3 \cdot 35 l (\sqrt{\text{H}^3} - \sqrt{h^3}),$

when H and h are measured in feet, and the quantity discharged is measured in cubic feet per second; but when, as in the present instance, H and h are measured in inches, and the quantity is desired to be stated in cubic feet per minute, $3 \cdot 35$ must be multiplied into 60, and divided by the square root of the cube of 12, which is $41 \cdot 56$. Under these circumstances, therefore, the quantity discharged in cubic feet per minute is stated as

$Q = 4 \cdot 83 l (\sqrt{\text{H}^3} - \sqrt{h^3}).$

The depth H, then, in the first experiment with the wheel No 1, being $11\frac{1}{4}$in., and the depth h being 4in.,

$Q = 4 \cdot 83 \times 11 \cdot 30 \times (\sqrt{(11 \cdot 25)^3} - \sqrt{4^3}) = 1,621.$

In the same experiment, with No. 2 wheel, the depth H was $12\frac{1}{4}$in., and h $4\frac{1}{4}$ in.

$Q = 4 \cdot 83 \times 7 \cdot 90 \times (\sqrt{(12 \cdot 25)^3} - \sqrt{(4 \cdot 25)^3} = 1,301.$

The two quantities together are 2,922 cubic feet per minute, having a total fall of 6ft., and $2,922 \times 6 = 17,532$ cubic feet per minute falling 1ft.

There being seven pairs of stones running, with the wheat-dressing and flour-dressing machines, and the elevator and hoisting tackle, the quantity of water expended per minute per foot of fall was 2,505 cubic feet per pair of stones driven. If this be represented in horse-power it would be $\frac{2,505}{528} = 4 \cdot 74.$

In the second experiment with No. 1 wheel, the height H was $10\frac{1}{2}$in., and h 3in.

$Q = 4 \cdot 83 \times 11 \cdot 30 (\sqrt{(10 \cdot 50)^3} - \sqrt{3^3}) = 1,572.$

In the same experiment, with No 2 wheel, H was $12\frac{1}{2}$in., and h $3\frac{1}{2}$in.

$Q = 4·83 \times 7·90 \left(\sqrt{(12·50)^3} - \sqrt{(3·50)^3}\right) = 1,437.$

These together are 3,009 cubic feet per minute, falling 6ft., equivalent to 18,054 cubic feet falling 1ft., and $\frac{18,054}{7} = 2,579$ cubic feet of water expended per minute per foot of fall per pair of stones driven.

In the third experiment with No. 1 wheel, H was $11\frac{1}{4}$in., and h $3\frac{1}{2}$ in.

$Q = 4·83 \times 11·30 \left(\sqrt{(11·25)^3} - \sqrt{(3·50)^3}\right) = 1,700.$

In the same experiment with No. 2 wheel, H was 12in., and h $3\frac{1}{2}$in.

$Q = 4·83 \times 7·90 \left(\sqrt{(12)^3} - \sqrt{(3·50)^3}\right) = 1,339.$

Together, these are 3,039 cubic feet per minute falling 6ft., or 18,234 falling 1ft., and $\frac{18,234}{7} = 2,605$ cubic feet of water per minute per foot of fall per pair of stones.

The mean quantity falling 6ft., derived from the three experiments, is $\frac{2,922 + 3,009 + 3,039}{3} = 2,990$ cubic feet per minute for seven pairs of stones, or 2,563 cubic feet per minute per foot of fall for each pair, and $\frac{2,563}{528} = 4·85$ horse-power.

With regard to the large quantity of water shown to be used per pair of stones with those undershot wheels, viz., 6,032, 6,060, 6,143, 6,224, 7,125, 6,128, cubic feet per minute per foot of fall, including the dressing machinery, the average being 6,285, if this be put into the form of horse-power it would be $\frac{6,285}{528} = 12$ horse-power nearly, per foot of fall per pair of stones. Every miller knows, or at least millers often say, that a pair of French burr wheat-stones of medium size can be driven by 4 horse-power, which would be $528 \times 4 = 2,112$ cubic feet of water per minute per foot of fall; and that is so when the water is used by way of its weight on overshot or high-breast wheels; and taking this as a standard the

HORSE-POWER PER PAIR OF STONES.

undershot wheels on which the above-named experiments were made do effective work only about one-third of the power expended, or more accurately $\frac{2,112}{62\cdot 85} = 33\cdot 6$ per cent. The owners of all those undershot wheels would be amply compensated for the expense of substituting for them the wheels called "Poncelet," which give an effective power of 55 per cent., and would use under the same circumstances $\frac{2,112}{\cdot 55} = 3,840$ cubic feet of water only, to do the same work, or with the same quantity of water would do more work in the ratio of 1·63 to 1, or 5 to 3. With regard to the diameter of the wheat-stones named, such as 4ft. 10in., 4ft. 8in., &c., these are rather larger than the medium size, which is perhaps 4ft. 4in. or so, but on the other hand the quantity of water named in each case was measured as fully as was consistent with my duty. The experiments with breast wheels show a much less quantity of water required per foot of fall per pair of stones driven, including the dressing machinery as before, viz., 2,833, 2,505, 2,579, 2,605, the average being 2,631 cubic feet per minute; and applying the same standard of 4 horse-power, or 2,112 cubic feet of water per minute falling 1ft., the effective power is $\frac{2,112}{26\cdot 31} = 80$ per cent. nearly.

SECTION XX.

WORK DONE BY WATER-WHEELS.

THE most direct measure of the work done by a water-wheel is the quantity of water pumped to a certain height by another quantity falling through a certain other height. The following instance affords such a measure. When I was an assistant to the late Mr. James Simpson (past President Inst.C.E.), I made the following trials for him with two low-breast wheels, each 19ft. diameter, No. 1 being 12ft. wide; and No. 2, 10ft. The fall of water was 6ft. Each wheel works three pumps. A cast-iron pipe, 10in. diameter, and 10,890ft. long, delivers the water into a service reservoir at a height of 148ft. above the level of the source from which the water is pumped. This reservoir, for a few feet in height at the top, has vertical walls, inclosing an area of 40,804 sq. ft. The trial consisted first of the quantity of water expended on the wheels, and secondly of the increase of the depth of water in the reservoir in a certain time, when the outlet valve was closed.

The gate of the wheel No. 1 is 11ft. wide, and it had a depth of water of 1·21ft.; the gate of No. 2 wheel is 9·75ft. wide, and it had a depth of water of ·96ft.

According to the construction and arrangement of these gates the proper co-efficient of discharge would probably be ·62. Without allowance for contraction of area or for diminution of velocity, the quantity would be $8·02 \sqrt{h} \times lh \times \frac{2}{3} = 5·35\ lh\ \sqrt{h}$, and if this be reduced, for the effects of contraction, &c., by the co-efficient ·62, the actual quantity would be $3\frac{1}{3}\ lh\ \sqrt{h} = 3\frac{1}{3}\ l\ \sqrt{h^3}$. Accordingly the quantity in cubic feet per second would be

$3\frac{1}{3}$ (11 $\sqrt{(1\cdot21)^3}$ + 9·75 $\sqrt{(\cdot96)^3}$ = 80, or 4,800 cubic feet per minute.

This would be the quantity falling from a still head of water; but the form of the channel is such that the water approaches the wheels with a velocity of 1½ft. per second, as observed by floats, and the height h must be increased by a height which is sufficient to produce this velocity, which is $h_a = \cdot 035\ v^2$. The formula, therefore, is $3\frac{1}{3}\ lh$ $\sqrt{h + h_a} = 3\frac{1}{3}$ (11 × 1·21 $\sqrt{1\cdot21 + 0\cdot35 \times 2\cdot25}$ + 9·75 × ·96 $\sqrt{\cdot95 + \cdot035 \times 2\cdot25}$) = 83 cubic feet per second, or 4,980 cubic feet per minute, which, multiplied into the fall, 6ft., and divided by 528, is equal to 56 horse-power. To compare this with the effect produced in 11 hours' pumping, the water surface of the reservoir was raised 19in., the quantity in that time amounting to 64,606 cubic feet, or at the rate of 1·63 cubic feet per second. The area of the 10in. pipe is ·545 sq. ft., and the velocity through it must, therefore, have been 3ft. per second. The head of water required to produce this velocity is

$$h = \frac{lv^2}{2,500d} = \frac{10,890 \times 9 \times 12}{2,500 \times 10} = 47 \text{ feet.}$$ The head against which the pumps lift, is, therefore, 148 + 47 = 195ft. If this be multiplied into the number of cubic feet per minute, 98, and divided by 528, the effect is that of 36 horse-power, or 64 per cent. of the power expended.

This is the utmost quantity of water the wheels would carry, purposely tried, and the velocity of the water through the pumping main was greater than it should have been for economical working. If a less quantity of water had been expended on the wheels, such as would have produced a velocity in the 10in. main of 2ft. per second, or thereabouts, the result would probably have been a greater effect than the percentage above named.

In the neighbourhood of Sheffield water power is much used for grinding saws and cutlery, and saw-grinding is very heavy work; but still heavier work is that of forging and grinding anvils, and for this kind of work water power is preferred to steam; the work is not very regular, and in the intervals the water accumulates and the great power required is exerted during short periods of time. A grinding establishment is called a wheel. On some of the streams there is a succession of very considerable falls of water, and, although the quantity is not large, a great deal of work is done by the same stream. One of them, for instance, has an ordinary flow of about 600 cubic feet per minute where it begins to be used, increasing to 1,000 or more where it falls into a larger river, and in this distance of a few miles there are twenty grinding wheels, mills and forges. The fall at No. 1 wheel is $18\frac{1}{2}$ft. The wheel is overshot, 14ft. diameter, 5ft. wide, with 5ft. head of water in the pentrough over it. There are 16 grindstones for table knives and files; but it is the trough in which the stone runs that the grinder claims as his possession; and this is a wheel of 16 troughs. When 14 of these are occupied, the sluice-gate or "shuttle" is drawn $2\frac{1}{2}$in. The length of the opening being 4ft. 4in., its area is ·91 sq. ft. The form of the opening is such that probably ·63 would be the proper co-efficient of discharge when the sectional area of the stream is taken to be that of the sluice-opening, and $8 \sqrt{h} \times ·63 = 5·04$, or say, $5 \sqrt{h}$, for the mean velocity of all the particles of water passing through the opening, when resolved in a direction at right angles to it, or in the direction which the successive central particles take; and the quantity discharged would, therefore, be $·91 \times 5 \sqrt{5} \times 60 = 606$ cubic feet per minute.

Wheel No. 2 is 15ft. diameter, 6·58ft. wide, with 4ft. head of water in the pentrough. There are 8 grindstones for table knives, and 10 for "light" work. All these being occupied, the shuttle is drawn 2in. The length of opening is 6·17ft., and the quantity of water discharged would be

$$6·17 \times \frac{2}{12} \times 5 \sqrt{4} \times 60 = 618 \text{ cubic feet per minute.}$$

The fall is 18ft.

No. 3 is an overshot-wheel, 11·75ft. diameter, 7·25ft.

wide, with a head of 3ft. in the pentrough at the time of the trial to ascertain the quantity of water. There are 9 table-knife and 4 razor stones. With these running, the shuttle is drawn 2½in. The length of opening is 6·85ft., and the area 1·42 sq. ft. The quantity of water is $1 \cdot 42 \times 5 \sqrt{3} \times 60 = 736$ cubic feet per minute. The fall is 14ft.

No. 4 has two water-wheels, each 11ft. diameter, one of them 5ft. wide, the other 4·67ft. The fall is 15ft. When the trial was made there were running 10 table-knife stones, and 4 razor stones. The sluice-openings are respectively 4·75ft. drawn 1½in., and 4·50ft., drawn 1¼in., discharging together 656 cubic feet per minute.

These are all overshot-wheels, with pentroughs, from near the bottom of which the water is discharged upon the wheel, and in which the head varies from time to time, according to the season, or according to the work being done; whether it is more or less than the power of the stream for the time being.

The next one is a wheel of a different kind.

No. 5 is a high-breast wheel, 15½ft. diameter, 8½ft. wide. The fall is 13ft. with a full head of water; but it varies at different times, for the same reasons as those before stated. For this wheel, however, the whole fall of water, whatever it be at various times, is made use of without loss of head by letting the water fall over the top of a movable lip, which is raised or lowered at pleasure, so as to discharge the water into the buckets of the wheel at the highest possible level. The circumference of the wheel moves closely under a series of cast-iron bars arranged across the face of the wheel, 1¼in. apart, the clear length of waterway of the openings being 6ft. 9in. (see Fig. 39, p. 176).

When the trial was made, there were 11 saw-grinding stones running, and five of these openings were exposed, the head upon each successive bar from the top increasing by 2in.

Thus, upon the first opening the head was 2in.; upon

the second, 4in.; upon the third, 6in.; fourth, 8in.; and fifth, 10in. The quantity of water passing through these openings may be estimated by the velocity due to the head upon each, which will be proportional to the square root of the head in each case.

The passages between the bars partake of the character of a tube projecting inwards through the side of a tank into the body of water—a form which does not facilitate the passage of the water through it, but retards it, as compared with a tube having a rounded mouth, or even with a tube having sharp arrises, if they form part of the side of the vessel. On the other hand, the length of each passage is short, being about 4in. On the whole, it is considered that a proper co-efficient of discharge is ·7, and that the quantity in cubic feet per second would be expressed by

$$Q = A \times 8 \sqrt{h} \times \cdot 7.$$

The expression $\sqrt{h}$ must be the sum of the square roots of all the heights, in feet, thus:

$$\sqrt{\cdot 17} + \sqrt{\cdot 33} + \sqrt{\cdot 5} + \sqrt{\cdot 66} + \sqrt{\cdot 83} = 3 \cdot 41$$

and

$$Q = 6 \cdot 75 \times \frac{1 \cdot 25}{12} \times 5 \cdot 6 \times 3 \cdot 41 \times 60 = 798 \text{ cubic feet}$$

per minute. This method of applying water to a wheel may be either by means of a gate sliding in the arc of a circle by means of rack and pinion, or it may be by means of a roll of leather or other web, covering the openings and folding downwards, its position at any time corresponding with the top of the sliding gate in the other method. The roll has less friction, and is in some other respects preferable; but the gate is, perhaps, the more lasting.

The same kind of work on another kind of wheel will exhibit the difference in effect between a high-breast wheel upon which the water is laid without loss of head, and upon which it acts with its full weight, or nearly so, and a wheel of less diameter than the fall of water, working under a pentrough.

No 6 has two water-wheels, one 9ft. diameter, 7ft. wide, with a head of water in the pentrough of 2·42ft.; the other wheel is 8·67 diameter, 6·50ft. wide, with the same head of water. The sluice-opening in the first-named wheel is 6·75ft., and in the other 6·25ft., and the shuttle was drawn 2in. in the one and 2½in. in the other, and the quantity of water expended on both wheels was 1,080 cubic feet per minute. The fall is 11ft.

There were running 6 saw-grinding stones, and 2 scythe-stones. If the quantity of water expended on the high-breast wheel—viz. 798 cubic feet per minute—be multiplied into its fall, 13ft., and divided by the number of stones running, 11, the result is 943 cubic feet per minute per stone. If the same process be followed in the second case, the quantity of water is 1,080 cubic feet per minute, the fall 11ft., the number of stones running, 8, and the result is $\frac{1,080 \times 11}{8} = 1,485$ cubic feet per minute per foot of fall per stone running.

The tilt-hammer is almost an indispensable piece of machinery. One of the water-wheels works two of these (one at once), and a blowing-machine. The wheel is 9ft. diameter, 7ft. wide. The sluice-opening is $6·25 \times ·18$ft. $= 1·12$ sq. ft. The head upon it is 1·92ft.

The quantity of water discharged when one tilt-hammer and the blowing-machine are in work may be estimated at $1·12 \times 5 \sqrt{1·92} = 7·78$ cubic feet per second, or 467 cubic feet per minute. The fall is 11ft., and $467 \times 11 = 5,137$ cubic feet per foot of fall per minute. The hammer is 2½cwt., with a fall of 7in., and makes 250 strokes per minute.

At an anvil-forge the grinding-wheel is 9ft. diameter, 5·85ft. wide, overshot; head of water in pentrough, 3ft., sluice opened 1½in. Total fall of water, 12ft.

The wheel for the iron forge is 10ft. diameter, 4ft. wide, buckets 11in. deep, works one hammer with a sluice-opening of 4in., 4ft. in length, with a head of 3·10ft.

The shear-steel forge, which works one hammer, is a

breast-wheel 12ft. diameter, 3½ft. wide, having buckets 12in. deep. The sluice-opening is 3·40ft. long, and 6in. wide, with a head of 3·10ft.

The shearing scrap wheel is overshot, 9½ft. diameter, 4ft. 4in. wide. The sluice-opening is 3in. wide, 4ft. long, with a head of 3·10ft.

Another anvil forge has a breast-wheel 20ft. diameter, 7·17ft. wide. The fall is 13ft. When the following machinery is working, the sluice-opening is 1½in., the length being 7ft., and the head of water 3ft.—viz. blowing cylinders which supply 18 anvil hearths, 6 smiths' fires, 1 shear-steel furnace, and 1 iron furnace. With the following machinery added, the opening of the sluice is 2in.—viz. 2 slide lathes, 1 planing machine, and 1 drilling machine.

The anvil-grinding wheel is overshot, 11·50ft. diameter, 6ft. wide. The sluice-opening 5·60ft. long, 1½in. wide, the head of water being 3ft.

Coming lower down the streams into the main rivers, the wheels become undershot with larger quantities of water and less height of fall.

The first example of a forge with this kind of wheel—that is undershot, with the water confined in a close-fitting race—is a wheel 14ft. diameter, 6ft. wide, the centre of which is ·42ft. above the level of a full head of water. The head at the time the following trial was made was 1ft. below the full head. There are two forge-hammers, one 5cwt. with 16in. fall, the other 4cwt. with 14in. fall. The present fall of water is 5ft. The sluice is 6ft. wide, and when the 4cwt. hammer is working it is drawn 10in.

Another wheel works two tilt-hammers of 2½cwt. (one at once), with a fall of 7in., and one forge-hammer of 4cwt., with a fall of 12in. The diameter of the wheel is 14ft., width 5ft. 9in. With the present head of water, 1ft. below full head, the sluice-gate is drawn 9in. to work one tilt-hammer. The wheel of another tilting forge is 13ft. diameter, 5ft. 9in. wide, the centre being ·42ft. above the level of a full head of water. There are two tilt-hammers of 2½cwt., with a fall of 7 in., making 300 strokes per minute.

At the next works of this kind the wheel is 13½ft. diameter, 4ft. wide, the bottom of the sluice 1·42 feet below the present head; sluice-opening 4·50ft. wide; gate drawn 8½in. for one tilt, and 16½in. for the forge-hammer, which is 3½cwt., with 16in. fall, the tilts being each 2cwt., with 7in. fall.

Another kind of work is that of a wire-mill. The machinery consists of two wire-blocks 20in. diameter, and one 24in., and two fans; also of six grinding-stones for hackle-pins, 2ft. 9in. diameter. The wheel is overshot, 10·68ft. diameter, 4·5ft. wide. The sluice-opening is 4ft., and is drawn 3in., under a head in the pentrough of 3·75ft. The quantity of water expended would be—
$4 \times ·25 \times 5 \sqrt{(3·75)} \times 60 = 582$ cubic feet per minute. The fall of water is 15ft.

SECTION XXI.

TURBINES.

WHEN the height from which water falls exceeds about 40ft. it cannot be economically applied by way of its weight acting with a moderately slow motion on a single vertical wheel of ordinary construction, although a fall of 80ft. or 90ft. may be utilised on two vertical wheels placed one above the other; not so well, however, directly the one above the other, as when placed sideways in the manner shown in Figs. 43 and 44. In this arrangement, both wheels move in the same direction; but if one be directly over the other, they move in opposite directions. Moreover, it is sometimes desirable to work one wheel without the other, and in that case also this arrangement is convenient, and especially so when there is a side stream, high enough to be conducted into a tank over the lower wheel, but not into the higher one. The upper wheel shown in the diagram is 50ft. diameter, and 6ft. wide, the lower one 48ft. diameter and 7½ft. wide, both of wrought iron. In this instance the upper wheel is supplied by an 18in. pipe from a reservoir on the main stream, and by an arrangement of the sluices and stop-cocks the water is turned on to either wheel from the main source.

But when the fall of water exceeds about 30ft., it is more economically applied by way of its free velocity on a smaller wheel, usually placed horizontally, whereby the force of the water is diffused over the whole circumference, requiring, therefore, a smaller diameter; and this kind of wheel is suitable to any height of fall, say, from 3ft. to 300ft., and to various heights in the same place, for it works in backwater as well as free, and makes effective whatever fall there may be for the time being. This is an advantage where the quantity of water varies greatly

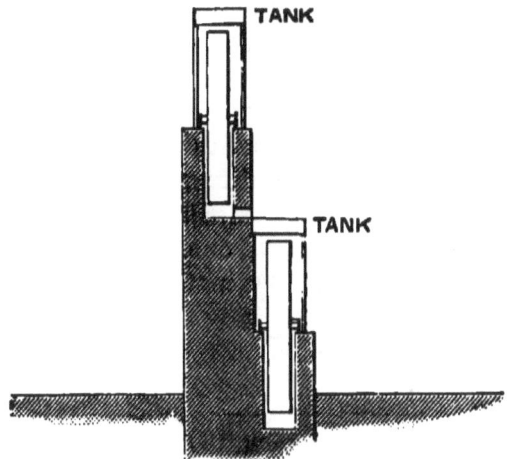

Fig. 43.

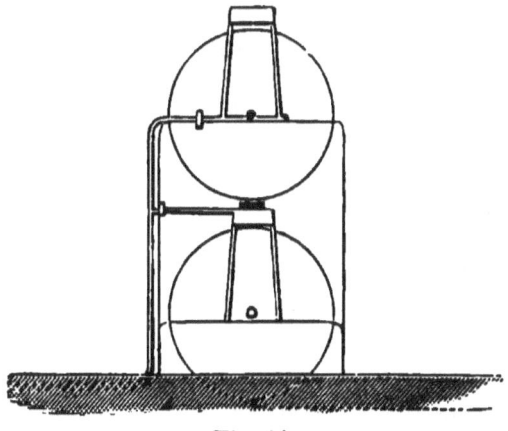

Fig. 44.

at different times, as upon a stream where the flood-waters are not impounded. The force brought to bear upon these wheels is of two kinds—direct pressure and reaction, the reaction being obtained by the liberation of pressure on one side of an arm, or of a cell or bucket.

The fundamental principle of the reaction-wheel is the release of pressure on one side of a revolving arm, whereby it is driven in the opposite direction with a pressure equal to that taken away. Fluid pressure acts in opposite directions at the same time, with equal intensity in both directions; and when the resistances are less than its pressure, it also moves in opposite directions at the same time, the velocity in either direction being inversely as the resistance; and if the abstraction of the quantity of water, due to the motion in one or both directions, be replenished at the head, the motion will be continuous under the same head.

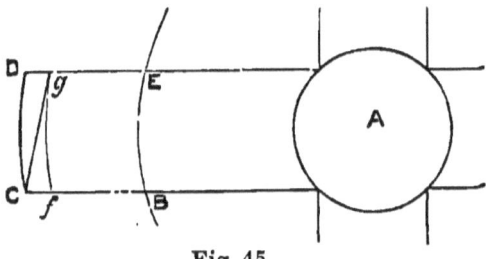

Fig. 45.

If the vertical pipe or chamber A, in Fig. 45, communicate with the head-water, and an arm, or two or more arms, project from it; or if passages be made between it and a wheel revolving round it, one of the cells or buckets of which may be represented by the two opposite vanes, B C and D E, with the closed end, C D, the water presses with equal force on the two sides, they being of equal area, and there is no motion. Neither would there be any motion if the end were closed between C and g, instead of between C and D, for the pressure upon the end C g in the direction of the tangents at every point between the two vanes is equal to the pressure upon D g only; and the pressures would equally be balanced whatever form, whether straight or curved, be given to the vane C g E.

But if an opening be made in one side, as from C to f,

that side of the cell would be relieved of the amount of pressure due to the area of the opening, the pressure which it had before being now transferred to the issuing water, and forcing it out with the velocity due to the head, and the obstructions to its flow. If there were no such obstructions the velocity in feet per second would be $\sqrt{64h} = 8\sqrt{h}$, h being the head of water in feet; but as there are obstructions in every form of channel, and however they may be lessened by good forms, the velocity of issue from turbine-orifices will probably in no case exceed $7·5\sqrt{h}$, considered as that of water issuing from an orifice which has no motion of its own. Such an opening being made, the vane opposite to it has upon it the pressure due to the head of water, and would have to be held by an external force equal to that which presses out the water with the velocity $7·5\sqrt{h}$.

If, at the same time, an opening were made from D to g, the water would issue with the same velocity, and if the two openings, D g and C f, were of equal area, the same amount of pressure would be taken away from each side, and the wheel would remain at rest. To prevent the issue of water from the opening D g the wheel would have to be turned by an external force acting in that direction with a velocity equal to that with which the water issued when the wheel was at rest, and when that orifice attained this velocity there would be no issue from it, because there would be no pressure upon it; but there would be twice the pressure upon the opposite one, C f, and the water would there issue with twice the velocity due to the head when the orifice was at rest; so that when the wheel moves in one direction with the velocity due to the head of water, and the water issues from the orifice in the opposite direction with the same velocity, the virtual velocity of the water is twice that due to the head. If, for instance, the head were 16ft., the velocity would be $7·5\sqrt{16} = 30$ft. per second—assuming it for the moment to be the utmost possible; but if, at the same time, the orifice itself be moved in the opposite direction with the same velocity, its motion must be added to that due to the head upon the orifice, producing in the issuing water a virtual velocity of 60ft. per second.

The inherent power of a head of water, however, could not produce and maintain continuously this double velocity by its steady pressure merely. All it could do continuously would be to produce a velocity through the orifice in one direction of 30ft. per second, and 15ft. in the wheel carrying the orifice in the opposite direction. This pressure, which call p, is released from the wheel at Cf, while it continues to act upon Dg. But although this pressure exists upon Dg when the wheel begins to move, the head of water cannot maintain it with any velocity. The pressure begins to diminish when the wheel begins to move, until, if it were to move with the velocity due to the head of water, which call v, there would be no pressure at all upon the vane. Between these two extremes, of pressure on the one hand and velocity on the other, there is a certain mean which can be maintained by the head of water, and under which condition the momentum will be the greatest possible, that is when half the maximum pressure, or $\frac{1}{2} p$, acts with half the maximum velocity, or $\frac{1}{2} v$, the momentum then being $\frac{pv}{4}$.

When the velocity of issue from Cf is 30ft. per second in a state of rest, and the wheel moves with a velocity of 15ft. per second in the opposite direction, the virtual velocity of the water through the orifice is 45ft. per second, two-thirds being due to the direct pressure upon the orifice, and one-third to the reaction of the opposite vane upon it, and the quantity of water expended is increased in the same proportion. If, for the sake of symmetry and uniformity of motion, a corresponding opening in the opposite direction be made in an arm on the opposite side of the supply pipe A, the combined areas of the openings multiplied into the velocity of the water would represent the quantity of water expended.

But no account is here taken of the centrifugal force which acts upon the water, and upon the vanes of the wheel through its medium. It would appear to be equal in effect to the force of reaction, for reaction wheels move with a velocity twice that due to the head of water, when perfectly free to move in a direction opposite to that from which the water issues. The original form in which the principle of reaction of hydraulic pressure was applied to wheels was that of two straight arms attached to a central vertical pipe, the whole revolving together. That was the invention of Dr. Barker. Mr. Whitelaw adopted the same principle on larger wheels, varying the forms of the arms to spiral curves, as shown in Fig. 46, and making them revolve round the central pipe instead of being attached to it, and

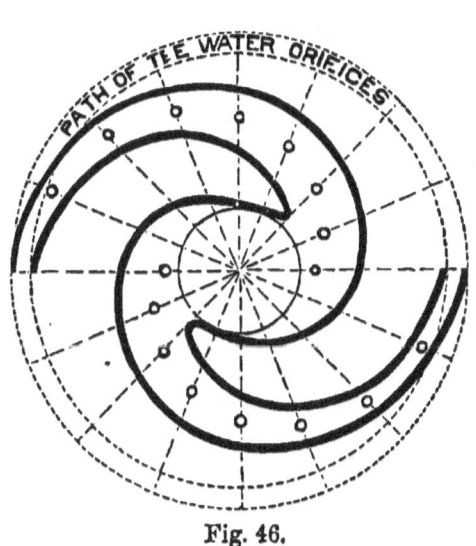

Fig. 46.

also relieved the bearing of the upright shaft from the weight of the water, by introducing it underneath the arms. Mr. Whitelaw made many experiments with Barker's straight arms, curved merely at the ends in the direction of the flow of water, introducing the water, however, underneath. The object of the experiments was to ascertain the percentage of useful effect of this kind of wheel. One of these experiments shows the velocity of the wheel, taken at the centres of the orifices, to have been 38ft. per second when the wheel was unloaded. Taking the stated quantity of water expended in a minute, and the given areas of the orifices, the virtual velocity of the water must have been 40ft. per second.

In this experiment the wheel was running free. The water issued with the velocity 40 − 38 = 2ft. per second, being the difference between the virtual velocity of the water as deduced from the quantity given and the observed velocity of the wheel; and it may be assumed that if in this experiment the wheel had run perfectly free, the velocity would have been 40ft. per second, or twice that due to the head of water. But there is the important difference that the water would not, in fact, have issued with that velocity under the same head from an orifice at rest. The head was 8ft., and at the utmost the velocity from an orifice at rest would not have been more than $7 \cdot 5 \sqrt{8}$ = 21ft. per second. But it could not have been in this case so much. The virtual velocity of the water is found to have been 40ft. per second, which was a double velocity, the wheel running free, and without resistance; that is, it would have had a speed of 40ft. per second if there had been no resistance at all, instead of the 38ft. which it had in fact. One-half of this velocity, or 20ft. per second, is that due to the head of water, the issue taking place from an orifice at rest. Dividing this velocity by the square root of the head, or $\frac{20}{2 \cdot 83}$ the co-efficient is found to be $7 \cdot 07$, and it agrees with those which have been found by experiments under other circumstances on analogous forms of opening.

In respect of a loaded wheel, the mean of 14 experiments on the percentage of useful effect was 74 per cent. The average velocity of the wheel in these fourteen experiments was $22 \cdot 5$ft. per second, and the virtual velocity of the water 30ft. per second, as deduced in the manner before stated. The difference, or $7 \cdot 5$ft. per second, would be the actual velocity of issue from the orifices.

With water stored, there are many businesses which might be carried on within a few miles of the reservoirs by means of turbines; industries created, as it were, and where all engaged in them might live in an atmosphere free from smoke. The economic conditions of trade and employment are, on the whole, far too large a subject to be entered upon here; but it may at least be said that, so far as they go, and their capabilities can be developed, these water sources of power form sound and firm bases of industry, whether of new ones or of transfers from overcrowded localities, which have no other source of power than coal.

Wherever a turbine may be used, the two chief conditions which will determine its dimensions and power are the quantity of water per second and its pressure. The quantity per day of 24 hours will be, probably, the one directly derivable from the data. In some instances the night water may be stored where it is used, in others it may better be stored in the main reservoir, and given out during 12 or 14 hours a day. However that may be in different cases, the quantity here to be referred to is that due to the working time per second.

A turbine, unlike a vertical water-wheel, takes its supply of water at the end instead of at the beginning of the fall. The end may be immediately under the head-water, with no greater length of supply pipe than the height of the fall, in which case the pressure will be very nearly that due to the whole height; or the turbine may be at a considerable distance from the head-water. In the latter case, the available pressure will be reduced by as much as is due to the head of water required to deliver the given quantity through the pipe, according to its length and diameter, taking into account also its degree of straightness, for frequent and abrupt changes of direction in the flow of water materially increase that portion of the head required to deliver a given quantity at the end of the pipe, leaving so much the less available as useful pressure.

Whatever be the necessary deductions from the "fall" on these accounts, the available head under which the turbine will work will be that remaining after these deductions have been made, constituting the immediate head-water, whether it remain in the closed pipe, or be delivered into an open reservoir or chamber immediately above the turbine. In the latter case the height between the two surfaces of the water above and below the turbine is the working head, and is, in all cases, the "head" here referred to; and as pressure is proportionate to the height, it will be considered as so many feet of water in vertical height. The velocity with which water under a given head issues from an opening is the same, whatever be the size of the opening. It would be the same as that of any heavy body falling freely through the same height, but that the necessary changes of direction which it must undergo in passing through outward-flow turbines retard its velocity by about one-tenth part. Through any form of opening whatever, that is to say, through one formed in the most favourable way for the free issue of the water, the velocity is retarded by about one-fortieth part, and it is this actual velocity, and not one due to a freely-falling body, which should be taken as the standard with which to compare the experimental velocities through turbines of different kinds. The velocity in feet per second of a body falling freely from a height h, in feet, is $8\sqrt{h}$; but for water, under the circumstances just referred to, it is $7\cdot 8\sqrt{h}$. These forms of opening, however, are not those practically made, and through these the retardation of velocity is about one-sixteenth part, which would reduce it to $7\cdot 5\sqrt{h}$. The forms of these openings approximate to one through which the water would flow without contraction, and are, perhaps, as near to that form as can be adopted in practice. But there are, in outward-flow turbines, further reductions to be made, which make the whole amount about one-tenth of the velocity of a freely-falling body, and the velocity is in this case expressed by $7\cdot 2\sqrt{h}$. It is usual to assume, as a standard of comparison, that the theoretical velocity of water, issuing from an opening under a given head, is the same as that of any heavy body falling freely through the same height, which would be $8\sqrt{h}$, or more properly $\sqrt{2gh}$ where g is the force of gravity $= 32$. This being assumed, the velocity represented by $7\cdot 8\sqrt{h}$ would be that due to a head 5 per cent. less than the actual head when the formula $\sqrt{2gh} = v$ is adopted, and it is called loss of head. By the same standard $7\cdot 2\sqrt{h}$ would imply a loss of head of 19 per cent., or 14 per cent. more than the "loss" in the first instance, which, however, hardly seems properly to be called a loss, if it never existed.

But by the usual standard the loss of head due to the hydraulic resistances of outward-flow turbines seems to vary from 14 to 20 per cent., and in taking 19 per cent. it would seem that a sufficient allowance is made for all the hydraulic resistances combined; and in this is included also the leakage between the fixed and movable parts of the turbine. The turbine consists essentially of a wheel with a hollow rim, across which are placed, at certain distances apart, curved vanes to receive the pressure of the water on their concave sides, and the angle at which they are placed across the rim varies accordingly as the water acts chiefly by direct pressure or by the pressure of reaction. In the outward-flow turbine the water issues horizontally all round the bottom of a cast-iron pipe or chamber, in which guide-blades are set, so as to conduct the water into the rim of the wheel at the proper angle, and it escapes through the outer circumference of the wheel. The rim is connected across underneath to the central upright shaft, which drives the machinery, by a strong disc of cast iron, the supply-pipe, or chamber, and the fixed guide-blades being suspended from a frame above, and hanging within the revolving wheel, and clear of the disc beneath, by which means (the method of M. Fontaine) the weight of the water and the chamber is taken off the bearings, and the shaft-friction much reduced. The revolving upright shaft is kept from the water through which it passes by being inclosed in a pipe, which supports the whole machine. For low falls of water the top of the supply-chamber is open, but for high falls the top is closed, and the shaft passes through it. In one case the water is delivered into an open reservoir or chamber immediately above the turbine, in the other the turbine works at the end of a closed pipe, which is under pressure all the way from the head-water, which may be at a considerable distance from the position of the turbine.

The available head of water, however, which produces the velocity before mentioned, is almost as easily ascertained as if it were immediately above the turbine. With the reservations named above, nothing can alter the velocity with which water issues from an opening under a given head. If the area of the opening be too large in proportion to the quantity of water, the head will fall; if it be not proportionately large enough, the head will rise, until an equality is established; it will rise, that is to say, if confined in a pipe, and it will of necessity either rise or spread out laterally. The main object is to transmit to the wheel as much as possible of the motion obtained. The driving power of the turbine is measured by the difference between the pressure of the water due to the head and its pressure at the point of egress. The motion of the water is an absolute quantity, to which that of egress and that of the wheel are together equal.

If pressure on the egress orifices of the wheel be wholly taken away, as it is when the orifices move with a velocity twice that due to the head of water, considered as the velocity due to that head when the orifice is at rest, then the whole motion is transferred to the wheel; but if the water issue from the wheel with any velocity—that is, if the velocity of the water be greater than that of the wheel—the pressure causing that velocity of issue is to be deducted from the pressure due to the head, before the useful effect of the power can be found. If, for instance, the velocity of egress were one-fourth of that due to the head, or $1\cdot 8\sqrt{h}$, the loss of head on this account would be $6\frac{1}{4}$ per cent. If the velocity of egress be a little less than one-fourth, so that the loss of head be exactly 6 per cent., there would be, with the 19 per cent. before mentioned,

25 per cent. of the power expended before any work was done. But there is yet another source of loss of power, viz., the friction of the upright shaft in its bearings, for which, perhaps, 2 per cent. may be added, making in all 27 per cent. of the power expended, the useful effect of the turbine being in this case 73 per cent. But this is not so much as has been found to be the useful effect of many outward flow turbines, both in this country and abroad.

In one of the "Abstracts of Papers in Foreign Transactions, &c.," of the Institution of Civil Engineers, is an abstract of an investigation into the maximum efficiency practically attainable in the three chief kinds of turbines —viz., outward flow, parallel flow, and inward flow, by J. C. Bernhard Lehmann, from experiments with thirty-six turbines of all sizes, from 1 to 500 h.p., made in order to ascertain the shaft friction, from which it appears that the percentage of total available power lost by hydraulic resistances was 14 per cent.; by unutilised energy carried away in the issuing water, 7 per cent.; shaft friction, both in the foot-step and bearings, 2 per cent., or a total loss of 23 per cent., and therefore a possible efficiency of 77 per cent. In another "abstract" of a paper by Prof. Richelmy, it is estimated, from many experiments, that the actual amount of the loss of head in the supply chamber, before the water has issued into the wheel, is between five hundredths and eight hundredths of the head, according to circumstances in different cases. It is assumed that the form of the opening leading from the central supply chamber to the wheel revolving around it is the most favourable for the free issue of the water, and that it issues from the supply chamber into the wheel cells without contraction; and, therefore, the diminution ought to be attributed to a lesser velocity of efflux, due no longer to the head, but only to this multiplied by the square of ·92 or of ·95, or by the square of a number intermediate between these two, which is a loss in the supply pipe or chamber of from 10 to 15 per cent. of the head and of the energy of the water, and is the same form of loss as that which has been stated as 12 per cent. of the head, the actual velocity being $7\cdot 5 \sqrt{h}$.

Fig. 47 is a plan, and Fig. 48 a section of the essential parts of an outward-flow turbine. One of the principal points to be considered is the working velocity of the wheel. Mr. Cullen, an Irish millwright and engineer, states in his treatise on the turbine that he has found by experience that the most effective speed of the inner circumference of the wheel is $4\cdot 4 \sqrt{h}$, when the head does not exceed 38ft. For high falls his experience is that the cube root of the height should be taken and multiplied by $8\cdot 1$. Thus, the velocity of the inner circumference of the running wheel for high falls is $8\cdot 1\sqrt[3]{h}$. These velocities, Mr. Cullen says, produce the greatest amount of moving power, and are about two-thirds of the velocity of the water on its entering the wheel. According to Mr. Cullen's rules, the diameter A B in the sketch should be =

Fig. 47.

$\sqrt{\dfrac{Q}{\sqrt[3]{h}}} + \cdot 1$, where Q = the quantity of water in cubic feet per second, and h the head in feet. Thus, if Q = 40, and h = 30, the inner diameter of the wheel would be 3ft. 8in.

The number of buckets he gives as N = $3d + 28$, d being the inner diameter as already found. Then the width of the buckets, B D on the sketch, should be $\dfrac{55\,d}{N}$.

Applying these rules to the case stated, the number of buckets would be 39, and the width 5in., so that, adding 10in. to 3ft. 8in., the outer diameter would be 4ft. 6in. The height of the buckets would be 4in. The velocity at the inner circumference would be 24ft. per second, and at the outer circumference nearly 30ft. per second. Comparing these with the rules in Molesworth's pocket-book of engineering formulæ, there is a little difference. The inner diameter would be about the same, but the outer diameter would be 5ft., and the velocity at the outer circumference would be 36ft. per second.

Mr. D. K. Clark, in his 'Rules, Tables, and Data,' gives rules (p. 941) for the dimensions of outward-flow turbines, which agree nearly with Mr. Cullen's, except in the diameters, inner and outer. By Mr. Clark's rule for the

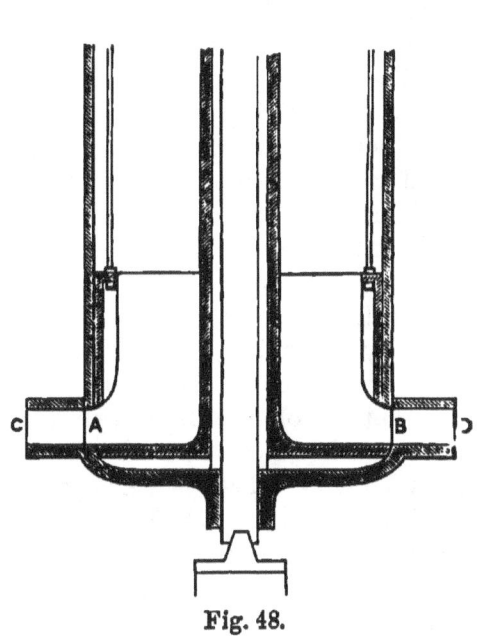

Fig. 48.

"exterior diameter," it would be, in the instance mentioned, $D = 4\cdot 85 \sqrt{\dfrac{P}{h\sqrt{h}}}$ as stated, P being the "actual horse-power of the turbine." Now, if in the above instance we take the actual horse-power of the turbine to be 75 per cent. of the power of the water, that is, 40 cubic feet per second falling 30ft., P would 102, for $\dfrac{40 \times 30}{8\cdot 8}$ $\times \cdot 75 = 102$, and $4\cdot 85 \sqrt{\dfrac{102}{30\sqrt{30}}} = 3\cdot 82$ft. for the "exterior diameter," which is only about 2in. more than the *interior* diameter should be, by Mr. Cullen's rule, and also Mr. Molesworth's. But if in Mr. Clark's rule we read D = *interior* diameter of wheel, instead of "exterior," the rule would agree nearly with the other two mentioned.

It is evident that water descending in a vertical pipe, or chamber, need not necessarily flow out of it horizontally, as in the outward-flow turbine, but that by placing the wheel across the upright pipe, or chamber, the water may continue to have its general direction vertical, and parallel with the axis, its direction for part of its descent only—that is, the part occupied by the depth of the wheel—being horizontal, or partaking of a horizontal and vertical motion combined, the water being finally delivered into the tail-race in a nearly vertical direction, on the underside of the wheel, being received into the buckets or cells on the top side of the wheel, and being guided into them by blades fixed in the bottom of the pipe or cylinder, and bent in a direction tangential to the circumference of the wheel. Thus the guide-blades at the bottom of the fixed cylinder occupy an annular space beneath which the buckets of the wheel revolve.

This kind of wheel is perhaps not quite properly to be called a turbine at all, but simply a horizontal wheel; nevertheless, it is convenient to classify it as a turbine, and to distinguish it as having a parallel flow. It is the principle on which are constructed those called Jonval, Fontaine, Henschel, Koecklin. Fig. 49 represents a development of part of the circle through the middle of the buckets of the wheel, vertically. On this line, the angle which the direction of the water, on its leaving the guide passages, makes with the horizontal face of the wheel is between 14° and 20°, varying accordingly as the water acts on the vanes by pressure or by reaction. In the former case the proper angle for the guiding passages of an axial turbine is stated to be 14° 30′ in a paper on the theory and construction of turbines, by Professor Fink, translated in the "Abstracts of

Papers in Foreign Transactions, &c.," of the "Minutes of Proceedings of the Institution of Civil Engineers, 1878." That angle is taken at the mean radius of the wheel, or at

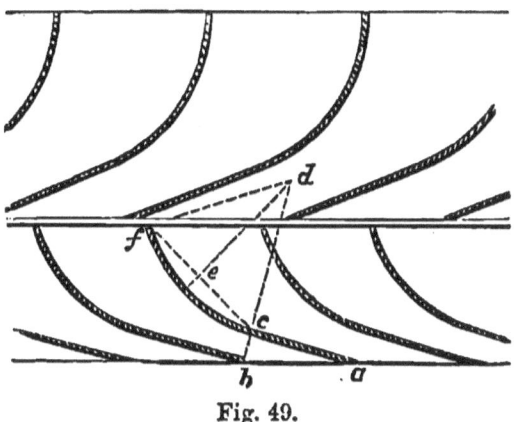

Fig. 49.

the distance from the axis which is a mean between the inner and outer radii; but at the circumference the angle is to be reduced to 12° 20'. In a reaction turbine the angle of the guide passages is 20° at the line of mean radius, and 17° 10' at the outer circumference.

In the diagram, the distance ab is the distance apart of the vanes of the wheel, which depends upon the number of buckets for any given diameter. According to the instructions of Prof. Fink, the number of buckets for pressure turbines should be 30 $\sqrt{D}$, taking D, the diameter of the wheel, in metres; and for reaction wheels the number should be 18 $\sqrt{D}$. These would be, if the diameters were taken in feet, $16\frac{2}{3} \sqrt{D}$, and 10 $\sqrt{D}$, respectively. Thus, for a reaction wheel 6ft. 8in. external diameter, the number of buckets might be 24, and the vanes 10½in. apart at the outer circumference of the wheel. A good proportion between the outer and inner diameters is said to be 1 to ·7; so that if the outer diameter were 6ft. 8in., the inner one would be 4ft. 8in., and the mean radius 2ft. 10in.

The curve of the wheel vanes may be drawn from Weisbach's instructions, given in Fairbairn's "Millwork," thus:—The distance ab in the diagram being the distance between the vanes, $= \dfrac{6 \cdot 29 \text{ radius}}{\text{Number of vanes}}$, which evidently gives the same result as before, being the circumference divided by the number of vanes. "The angle bac may be from 15° to 20°. Draw bcd at right angles to ac. Let A be the angle bac and B an angle taken arbitrarily at 100° to 110°. Lay off the angle dcf equal to $\dfrac{A + B}{2}$. Mark cf in the middle of its length, and draw ed at right angles to cf, cutting the other line in d, or lay off the angle dfc equal to the angle dcf. From the centre d draw the curve of the vane with radius dc or df, to which the straight part of the vane ac will be a tangent." Thus, if the angle B be taken at 105°, and A at 15°, the angle dcf would be $\dfrac{15 + 105}{2} = 60°$. In this case the angle cdf would also be 60°. The rule of Professor Fink for the two radii of the wheel is, for the outer one "$R = \cdot 85 \sqrt{\dfrac{Q}{\sqrt{h}}}$, Q being the quantity of water expended in cubic feet per second, and h the height of the fall in feet; and for the inner radius $r = \cdot 7$ R." Thus if the quantity of water be 48 cubic feet per second, and the fall 10ft., $R = \cdot 85 \sqrt{\dfrac{48}{\sqrt{10}}} = \cdot 85 \times 3 \cdot 9 = 3 \cdot 32$ft., or say 3ft. 4in., and the inner radius $= 3 \cdot 32 \times \cdot 7 = 2 \cdot 32$ft., or 2ft. 4in. The effective power of such a turbine would be 40 horsepower, for it may be taken that an efficient working power of 73 per cent. of the whole power of the fall of water would be produced by a good turbine, whichever form of construction be adopted, and $\dfrac{8 \cdot 8}{\cdot 73} = 12$ cubic feet per second, falling 1ft. to produce 1 effective horse-power, and $\frac{48 \times 10}{12} = 40$.

This kind of turbine is suitable for low falls, and the admission of the water to the guide passages is, in some, regulated by lowering into the mouths of the guide passages wedge-shaped stops, so as to partially close the mouths, and diminish the quantity of water entering the wheel. In others the quantity passing through the wheel is regulated by a throttle-valve across the cylinder below the wheel, and sometimes merely by a sluice at the head, admitting more or less water to the cylinder. But when regulated by a throttle-valve below the wheel, or by stops in the mouths of the guide passages, the operation can be made self-acting, and a constant head maintained, more easily than when the regulation is effected at the head.

If a tube be formed below the wheel, airtight, having its lower end under the surface of the tail water, the wheel may be placed at any height above the tail water within 28ft. or so; that is, at such a height that the pressure of the atmosphere on the surface of the tail water shall be so much in excess of the weight of water in the tube below the wheel as to preserve the continuity of the water-column between the tail water and the head water; for although the water may have passed through the wheel at a height of 28ft. or so above the tail water, yet its tendency to fall away from the water above the wheel, and so to produce a vacuum, brings into force the unbalanced pressure of the atmosphere upon the head water; and the effect upon the wheel is the same as if it were placed at the level of the tail water, and had the whole head of water above it. With such a draught-tube attached, this kind of wheel is suitable for moderately high falls.

The maximum efficiency practically attainable in parallel-flow turbines, after deducting the various forms and amounts of loss, is as follows, according to the experiments of Mr. Lehmann before referred to, viz.:—"Hydraulic resistances, including leakage, between the running wheel and the guide-blades, 12 per cent.; loss by energy carried away in the issuing water, 3 per cent.; shaft friction, both in the step and bearings, 3 per cent.; or a total loss of 18 per cent., leaving a possible efficiency of 82 per cent." But we do not know that the proportional dimensions and angles of direction of the guide passages and wheel vanes in the turbines experimented upon by Mr. Lehmann were the same as those above mentioned.

When the quantity of water being used is exactly that for which the dimensions of fixed openings of guide passages are made, then the best percentage of useful effect will be obtained; but when stops are introduced to regulate the quantity of water entering the wheel, whether they be placed in the mouths of the guide passages or at their ends, the flow of water does not take place in the most favourable direction, and the efficiency of a given quantity of water and height of fall is diminished. The efficiency of a Fontaine parallel-flow turbine when fully charged is, according to Mr. Clark's 'Rules, Tables, and Data,' 70 per cent.; but when the quantity of water is shut off to three-fourths by the sluice, the efficiency is only 57 per cent.; and the efficiency of a Jonval turbine, which is also a parallel-flow turbine, with a full charge is 72 per cent. For this wheel Mr. Clark gives as the best velocity for the outer circumference, 70 per cent. of that due to the fall, which would be $\cdot 7 \sqrt{2gh} = 5\cdot 6 \sqrt{h}$. For the Fontaine turbine he gives as the best velocity at the mean circumference of the wheel, 55 per cent. of that due to the height of the fall, which would be $\cdot 55 \sqrt{2gh} = 4\cdot 4 \sqrt{h}$. If the proportion of the lengths of the two circumferences, inner and outer, be taken as before at $\cdot 7$ to 1, and therefore the mean circumference as $\cdot 85$ to 1 of the outer one, the velocity of the outer circumference corresponding to $4\cdot 4 \sqrt{h}$ at the mean radius would be $5\cdot 17 \sqrt{h}$. For a Fourneyron turbine, which is an outward-flow turbine, the efficiency is given as 79 per cent. with a full supply of water; but only 24 per cent. when the ends of the guide passages are so reduced in sectional area as to pass but one-fourth of the full supply; and the best velocity for the inner circumference of the wheel is stated to be $4·49 \sqrt{h}$. Mr. Cullen, who treats of the outward-flow turbine in his treatise on turbines, recommends, from the results of his own practice, a velocity of $4·4 \sqrt{h}$ for the inner circumference where the fall does not exceed 38ft.; but for high falls $8·1 \sqrt[3]{h}$.

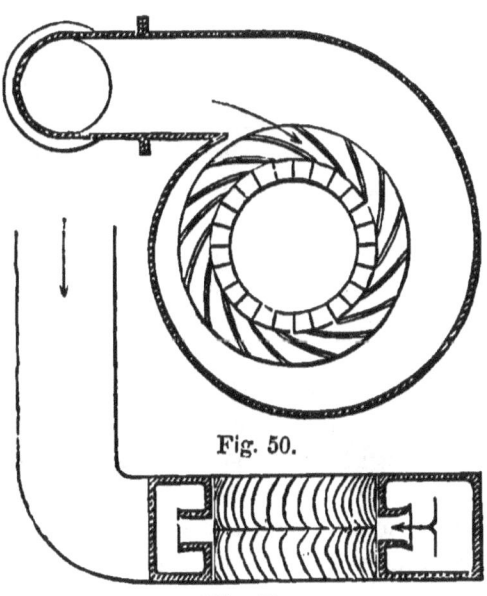

Fig. 50.

Fig. 51.

In another kind of turbine, the inward-flow, the water enters the wheel at its outer circumference. Fig. 50 is a plan, and Fig. 51 a vertical section of Schiele's turbine. It partakes of the character of both an inward and a parallel-flow, inasmuch as the water enters the wheel at its outer circumference, and is discharged upwards and downwards, parallel with the axis, or nearly. A whirling motion is given to the water before it enters the guide passages, by placing the wheel more to one side than the other of the casing inclosing the turbine, and making the form of the casing spiral. The direction thus given to the water from its entrance into the casing to its place of delivery into the wheel, being a gradual change of direction, obeys the law which should be observed in all water passages—that of avoiding abrupt changes of direction—and the water is delivered into the wheel with the least possible diminution of velocity and loss of head.

It may be observed that there is a division of the buckets near the middle of the depth of the wheel, and that there is a separate set of cells or buckets on each side of it, the water escaping upwards from one and downwards from the other; and by making the division a little below the middle of the depth of the wheel, more water is admitted above the division than below it, and the motion of the greater quantity being upwards, on its passage through the wheel, eases off the weight from the footstep, so that if the two divisions be properly proportioned there will be no pressure at all on the footstep. The difference between the upward and downward pressures of the water is, of course, transferred downwards to the fixed rim and its supports. It is convenient to do this to prevent wear of the footstep; but of course whatever power is required to ease off the weight from that bearing must come out of the power of the fall of water, and adds so much to the hydraulic resistances. Where the fall is considerable the wheel may be placed vertically, and the shaft lie horizontally, without intermediate gearing. In this arrangement, the water, as it leaves the wheel on both sides, is conducted by a pipe to the tail stream, and the wheel may be placed at any height above the tail stream, within about 30ft. Thus, in high falls, the water may be conducted both to and from the turbine in a cast-iron pipe. The foot of the pipe, which conveys the water away from the turbine, is turned horizontally, and in this short length of horizontal pipe a valve is placed to start and stop the turbine, worked by a rod from above.

In the *inward-flow* turbine, proper, the water enters at

its outer circumference, where the velocity of the wheel is greatest, and escapes near its centre, where it is least, and this accordance tends to prevent shock on the entrance of the water into the wheel. The general direction of the water, being towards the centre, is opposite to that in which the centrifugal force of the water acts. The height of the column of water which produces a pressure equal to that of the centrifugal force of a mass of water proceeding from itself is exactly that height which produces the velocity with which it moves, and the pressure produced by the centrifugal force and that which produces the velocity, taken together, are equal to that of the whole head of water.

If these two pressures—the pressure of the centrifugal force and that due to the velocity of the water—be unequal, the motion will not be uniform; it will attain uniformity only when the two pressures are equal to each other—that is, when the pressure of the centrifugal force is that of half the head of water, the other half producing the velocity with which the water issues; and the effect is the same as would take place if the issue of water from an orifice at a given depth below the surface had upon it on the opposite side a head of water equal to half the whole head. In the turbine, this half of the pressure is transferred as pressure upon the vanes of the wheel. The proper velocity of the outer circumference of an inward-flow turbine should, therefore, be determined on this principle—that is, it should be

$$\sqrt{\frac{2gh}{2}} = \sqrt{32h} = 5 \cdot 66 \sqrt{h}.$$

The vortex wheel of Professor James Thomson is an inward-flow turbine, and he first called attention to this principle of the equalisation of the two forces. In the vortex turbine the guide-blades which give the water the proper direction into the wheel are fixed when the quantity of water is constant; but when it is variable the guide-blades are made to turn on pivots near their inner ends, so as to vary the width of the openings which admit water to the wheel. Four of them are considered sufficient, and sometimes even a less number. Fig. 46 represents a plan of a wheel, showing a great number of vanes; but only alternate ones go through from the outer to the inner circumference. It is desirable to have many vanes in this and all other turbines, but at the same time not to bring them too near each other at any point. According to a published statement by Messrs. Williamson, of Kendal (the makers of these turbines), they have been applied to many different heights of fall, varying from 4ft. to 247ft.

Fig. 52.

The "centre vent" American turbine is on the principle of inward flow. Mr. Hett, of Brigg, who follows the same principle in his turbines, with his own improvements, in a published statement of the proper sizes for various heights of fall and quantities of water, appears to make the velocity of the outer circumference about $5 \cdot 56 \sqrt{h}$, taking an average of all diameters, and all heights of fall; but the velocity varies a little with both the diameter and the fall. Mr. Nell, of London, in a published statement of the same particulars concerning the "Victor" turbine, which is likewise on the inward-flow principle, and of American manufacture, appears to make the average velocity for all heights of fall and diameters of wheel about $5 \cdot 26 \sqrt{h}$. Mr. J. B. Francis, C.E., found, from experiments upon a central-vent turbine, that an efficiency of $\cdot 797$, or nearly 80 per cent., was obtained

when the velocity of the outer circumference was
$·64 \sqrt{2gh} = 5·14 \sqrt{h}$, and that this efficiency was also
obtained when the velocity was $·708 \sqrt{2gh} = 5·68 \sqrt{h}$.
The average velocity of all the high efficiencies of 78 to
79 per cent. of the power expended was $·65 \sqrt{2gh} = 5·22 \sqrt{h}$.

The explanation of the effect of the opposing centrifugal force, alluded to above, stated by Professor James Thomson, the inventor of the "vortex" wheel, is that "by the balancing of the contrary fluid pressures due to half the head of water, and to the centrifugal force of the water in the wheel, combined with the pressure due to the ejection of the water backwards from the inner ends of the vanes of the wheel when they are curved, only one-half of the work due to the fall is spent in communicating *vis viva* to the water, to be afterwards taken from it during its passage through the wheel; the remainder of the work being communicated through the fluid-pressure to the wheel, without any intermediate generation of *vis viva*. . . . . On any increase of the velocity of the wheel, the centrifugal force increases, and so checks the water supply; and on any diminution of the velocity of the wheel, diminishes, and so admits the water more freely."

The truth of this principle was supported by the late Professor Rankine as follows:—"The action of centrifugal force in the regulation of the pressure within the wheel is of the following kind: It is favourable to economy of power that the effective pressure, immediately after entering the wheel, should bear a certain definite proportion to the effective pressure in the supply chamber, not differing much in any case from one-half. The centrifugal force of the water, which whirls along with the vortex-wheel, tends to preserve at its circumference the very pressure which is most favourable to economy of power; and the centrifugal force of the two discs of water contained between the wheel and the two shields or covers of the wheel-chamber prevents that pressure from making the water leak out between the wheel and the casing."

Another kind is the "Swain" turbine, an American one, reported upon in the 'Journal' of the Franklin Institute, April, 1875, by Mr. J. B. Francis, C.E., who made 146 experiments with this wheel. The direction of the water in its passage through the wheel is first inwards and then downwards. The discharging edge of each bucket is vertical in the upper part, parallel with the axis; but the lower part is curved, with a radius equal to one-fifth of the diameter of the wheel.

A scientific account of the action of water on turbines was given by Professor W. C. Unwin, in a lecture on Water Motors, at the Institution of Civil Engineers in March, 1885.

SECTION XXII.

Domestic Water-supply.

Whether the gallon or the cubic foot is the better unit of measurement of small quantities of water depends to a certain extent on the purpose of its use, and it may be convenient sometimes to have the quantity in gallons because of the lesser magnitude of the unit; but for large quantities, such as the capacities of storage reservoirs, there is not so wide a difference in the respective magnitudes of the unit as to make one very much more convenient than the other. There is some advantage in reckoning the quantity for domestic purposes in gallons, because the supply can be subdivided into gallons for each of several uses. It is indeed usual to state the supply for all purposes in a town in gallons per day per head of the population, as, for instance, 20, 30, or 40 gallons, instead of the corresponding quantities of 3·2, 4·8, or 6·4 cubic feet; but the number of gallons is a derived quantity, not an original one. Reservoirs and gauges, measured in feet, have shown that a certain number of cubic feet have been supplied to a town in a known period of time, the quantity per head varying in different towns with its trade and other circumstances; and when it is said that, for instance, 25 gallons per head per day are requisite, that is derived from the fact that for populations of similar requirements, and where the means of distribution are similar, the supply to the town has been at the average rate of 4 cubic feet per day per head of the population.

Assuming this to be an average supply, it may be

divided into the separate uses as follows. The different quantities are derived from various sources, and some of them modified from the originals, so that the statement becomes an estimate, merely, of approximate averages; but it is believed to be one, nevertheless, which is sufficiently near to have the value of an approximate estimate.

GENERAL DOMESTIC PURPOSES.

Gallons per head per day.

Drinking	$\frac{1}{4}$	
Cooking	$\frac{3}{4}$	
Washing—Persons	3	
" Clothes	3	10
" Utensils		
" Floors	3	
" Walls, doors, and windows		
" Paved yards		

EXCEPTIONAL DOMESTIC USE.

Water closets }
Baths .. .. } 100 where used; say, used by 1 in 20
Gardens .. } of the population; average 5

PARTLY DOMESTIC, PARTLY TRADES USES.

Offices
Animals
Carriages } 100 where used; say, used by 1 in 50
Brewing at home } of the population; average 2

TRADES USES.

Slaughterhouses
Breweries
Manufactures
Boilers of steam engines
Warehouses
Public buildings, including hospitals and workhouses
Public baths

PUBLIC USES.

Street watering
Flushing sewers
Fires
Fountains

Total 25

A domestic supply must, of course, be "pure and wholesome;" but with respect to the first of these requirements the qualification must be understood in a limited sense, for there is no "pure" water to be had on even a moderately large scale, either by collection from the surface and from minor springs in the upper parts of valleys, or from rivers in the middle and lower parts. In flowing off the surface, it gathers organic matter, and in running through the ground and issuing in springs it gathers mineral matter, and it often, in fact, contains both these; but it would be no advantage to collect it in the form of rain.

Dr. R. A. Smith remarks, in 'Air and Rain,' that the rain from the sea, in the western islands, contains chiefly common salt; that rain contains sulphates in larger proportion to the chlorides than is found in seawater; that the sulphates increase inland before large towns are reached; they seem to be a measure of the products of decomposition, the sulphuretted hydrogen from organic compounds being oxidised in the atmosphere; that the sulphates rise high in large towns because of the amount of sulphur in the coal used, as well as of decomposition; that as sulphuretted hydrogen and sulphide of ammonium oxidise in the atmosphere, the sulphates may be expected to increase in proportion to the amount of decomposing organic matter containing sulphur, such as albuminoid compounds; that when the sulphuric acid increases more rapidly than the ammonia, the rain becomes acid; that when the rain has so much acid that 2 or 3 grains are found in a gallon of the rain-water, or 40 parts in a million, there is no hope for vegetation in a climate such as we have in the northern parts of this country; but free acids are not found with certainty where combustion or manufactures are not the cause; and he does not mean to say that all rain is acid; it is often found with so much ammonia in it as to overcome the acidity. It seemed clear to him, from his experiments, that rain-water in town districts, even a few miles distant from the town, is not a pure water for drinking: and that if it could be got from the clouds in large quantities we must still resort to collecting it on the ground in order to get it pure; for the impurities of rain are completely removed by filtration through the soil: when that is done there is no longer any nauseous taste of soil or soot.

With respect to the wholesomeness of water, one question is whether hard or soft water is the more wholesome. There have been many contentions on this point. It has been said that soft water, such as is supplied to Manchester, Liverpool, Leeds, and generally the large towns in the manufacturing parts of the country, which are supplied by gravitation from the hills, is not wholesome because it does not contain lime, which is necessary to maintain strength of bone; and moreover that it becomes tainted with lead in passing through the house services, at least more than hard water does; but against this it is pointed out that it does not continue to dissolve the lead of the pipes, but forms by its chemical action a coating which thenceforward protects the lead from further action, and that the softer the water the sooner is this protective coating formed, after which there is no further action upon the lead.

The cases of lead-poisoning, which have undoubtedly occurred at various times from the action of rain-water upon the lead linings of cisterns in towns, cannot be attributed in a general way to the action of soft water upon lead. The house roofs from which such water is obtained furnish to the water, in large towns, acids derived from the atmosphere through which the rain falls, and these undoubtedly have an injurious action upon lead. Water stored for too long a time in reservoirs, large or small, would probably have the same effect, for at the bottom of such a reservoir there would be a considerable quantity of decayed and decaying organic matter, and this might communicate to the water a property which would have an injurious action upon lead; but all such objections as these are remediable, and their importance is far less than that of the numerous benefits of a water-supply which is not of too hard a character.

Water which is under 5 degrees of hardness by Dr. Clark's scale may be said to be soft water, and 15 degrees to be hard. The Chemical Commission of 1851 reported that "it may be useful to distinguish the quality known as 'hardness' of water according as it is of a temporary or permanent character. Perfectly pure or soft water, when exposed to contact with chalk (carbonate of lime), is capable of dissolving only a very minute quantity of that substance; 1 gallon of water, in weight equal to 70,000 grains, taking up no more than 2 grains of carbonate of lime. This earthy impregnation is said to give the water 2 degrees of hardness. But waters are often found containing a much larger quantity of carbonate of lime, such as 12, 16, or even 20 grains and upwards per gallon. In such cases the true solvent of the carbonate of lime, or at least of the excess over 2 grains, is carbonic acid gas, which is found to some extent in all natural waters. But this gas may be driven off by boiling the water, and the whole carbonate of lime then precipitates in consequence, or falls out of the water, with the exception of the 2 grains which are held in solution by the water itself. The gas-dissolved carbonate of lime gives therefore temporary hardness, curable by boiling the water. Other salts of lime, such as sulphate of lime, are generally dissolved in water without the intervention of carbonic acid gas, and therefore remain in solution, although the water is boiled, imparting hardness."

SECTION XXIII.

Service Reservoirs.

A SERVICE reservoir acts as a pressure regulator. It would be inconvenient to connect the service pipes of the distribution directly with the main pipe which conveys the water from the storage reservoir; it might be necessary to shut off the water at the reservoir for a day or two for repair of the conduit pipe or other purpose, and another inconvenience would be that the pressure due to the height of the column could not be economically made use of; the pressure would be least when most wanted. It is a common occurrence that during the busy hours of the day twice as much water per hour is used as the average hourly quantity, and at those times the velocity in the main would be doubled and the loss of head increased four times, and the serviceable pressure proportionately reduced.

The same thing occurs in the distributing main and service pipes in connection with a service reservoir, but with this difference—that, the velocity in them being less, there is less absolute loss of pressure; doubling the smaller velocity and squaring it for its effect does not produce much absolute loss of head. The excessive loss of pressure at the place where the water is to be used is avoided by interposing a service reservoir in which the flow of water expands over an area, and stands nearly at the same height at all times. This evenness of pressure is a practical advantage, although, no doubt, greater pressure would be obtained without the interposition of the service reservoir if but little water were being drawn from the pipes, and if the main conduit were a pipe laid below the hydraulic gradient. For any purpose for which the water would need to be shut off at the storage reservoir, such as the connection of a branch or the repair of the main, 48 hours would probably be time enough, for emptying the main, doing the work, and refilling the main, and in the following remarks it will be assumed that the service reservoir is to hold two days' supply of water.

A system has been in a few cases adopted of making very large service reservoirs, where the ground is sufficiently favourable, and thus increasing the storage capacity of the works materially, the form of the ground in these cases being such as to admit of water being impounded by an embankment on one or two sides of the site, or it may be partly on a third side; but usually there is little choice of situation for a service reservoir, its primary necessity being height, with reference on the one hand to the height of the highest ground to be supplied, and on the other to the height of the storage reservoir, the dimensions of the conduit and conduit pipe being made accordingly. The situation of a service reservoir being thus almost rigidly fixed, it is dug out of the ground for the most part, and the excavated earth embanked round it.

The four essential features of a supply of water by gravitation are the storage reservoir, the conduit, the service reservoir, and the distributing main; and service pipes. Filtration forms a fifth when it is necessary.

Where the supply of water in bulk may be undertaken in connection with the regulation of flood waters, the question of authority divides itself between the landowner on the one hand, and the local sanitary authority on the other, or it may be several local sanitary authorities in each case. The distribution of the water, and its filtration, if necessary, must be with the sanitary authority, but the service reservoir should be made by the reservoir authority as distinct from the distribution, authority, and the sanitary authority would take the water from the service reservoir at a price per 1,000 gallons or per million gallons. Near populous places, where there is usually excessive dust and smoke in the atmosphere, water for drinking cannot properly be exposed, and each sanitary authority would construct a covered service tank, into which the water would flow from the service reservoir. One service reservoir might suffice for several covered service tanks, which, in some cases, might belong to different local sanitary authorities, whether urban, as local boards, or rural, as boards of guardians. Whether one or several sanitary authorities take water from the service reservoir, each would lay a distributing main between the covered service tank and the service reservoir, and the service pipes in connection with

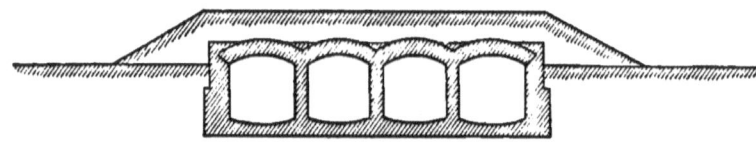

COVERED SERVICE RESERVOIR
Fig. 53.

it. Adjoining the service reservoir would be a measuring tank for each sanitary authority, in which would be the gauge, open to inspection, and so arranged as to give the quantity of water agreed upon from time to time up to a certain maximum.

The use of the service tank at the end of the distributing main would be to receive the excess of water not used, and to contain a day's supply in case of its being necessary to shut off the water at the service reservoir, of which notice could always be given so as to have the tank full. By the continued record of the state of the service tank it would be seen whether more water were being taken than was used, and it should be a condition of the agreement between the sanitary authority and the reservoir authority that the quantity may be varied from year to year as circumstances require, within a certain maximum. This, no doubt, would throw an onus upon the reservoir authority, but only a proper one, as being the dispenser of the water, and being able to estimate the proper provision to be made for present and prospective needs independently of any estimate made by the sanitary authorities.

Amongst the varying levels at which places to be supplied with water are situated in respect of any storage reservoir, three may be taken for example in which the covered service tanks, A, B, and C, would be at such levels that the highest would be about 100ft. below the lowest level to which the water in the reservoir would be drawn down, and if the population of the three places or districts be 30,000 in all, there would be required for a full supply 750,000 gallons a day.

OPEN SERVICE RESERVOIR
Fig. 54.

It would be a moderate view of the circumstances to assume that within an average distance of three miles of the three places a service reservoir might be made on ground sufficiently high to command them all, and that the water may be impounded at the height named within a distance of seven miles from the service reservoir.

The height of the service reservoir in such a case would be at about the mean height between the covered service tanks and the lowest water-level of the storage reservoir. There would then be about 16ft. or 17ft. fall per mile in the distributing main, and 5ft. per mile in the conduit. This would be sufficient if the water were carried along the hill-sides, but if any considerable length of low ground intervened, requiring to be crossed by an iron pipe, the total fall of the conduit could be a little increased and the fall of the distributing mains as much reduced.

An open reservoir should be at least 14ft. or 15ft. deep. The rays of the sun act through any shallow and stagnant water upon the bottom, and cause the growth of vegetation and insects. If the service reservoir in question were required to hold 1,500,000 gallons, or 240,000 cubic feet, being two days' supply, and it be made 15ft. deep, the water area at the mean depth would be 16,000 sq. ft., and if it were made square it would be, at that depth, 127ft. square. According to the nature of the ground, the slopes may be from $1\frac{1}{2}$ to 1 to 2 to 1. Let it be assumed that they are 2 to 1 throughout, both in the excavation and embankments, and that the top of these is 2ft. above the top-water level. The reservoir would then be at the top 165ft. square. The top of each bank may be 6ft. wide, except on the lower side, where, indeed, the width may be the same and the slope the same, but divided by a level bench of such width that the bulk of earth in the lower bank would, if continued in one slope, make the top width 9ft. Here we have no puddle trench, and to make the reservoir watertight, the puddle —if puddle be used—is continued across the floor and up the slopes of the excavation, and along the surface of the ground to meet and join with the puddle walls raised in the centre line of the embankments. The floor-puddle may be 2ft. thick if well made, and the wall-puddle should be at the bottom as thick as will allow the top to be finished 3ft. or 4ft. thick, with a gradual reduction in the thickness of 1ft. in each 6ft. in height. The surface of the floor-puddle and that upon the slopes, should be covered with broken stone or gravel, not less than 6in. in thickness, which may be continued up the slopes of the embankment to the top of the reservoir. This hardly forms of itself a sufficient protection of the puddle against the action of the water upon it, but if the interstices are filled with sand it does so. There is but little difference between this and concrete for this purpose. But to provide a surface which can be swept, brick-on-edge paving is added, or two courses of bricks laid flat.

If the water is not to be filtered, it is necessary that it should pass through fine wire-gauze strainers in going out of the reservoir. To hold these and the valves through which the water would pass, a structure of masonry is built.

The strainers are of copper wire, 40 to the inch; but as these, being of so fine a mesh, would too soon require removal if not protected, strainers of a mesh less fine, about 20 to the inch, are placed in front of them, and a double set of grooves is provided for these, so that before one set is removed the other may be inserted. The screens are removed and washed as occasion requires, being for that purpose and for strength made in small rectangular frames of wood, inserted in a larger frame which can be lifted out bodily, these being set one upon another to the height required.

The top soil, before the excavation is begun, would be taken off and reserved outside for soiling the slopes of the banks. The tops of the banks would be formed as footpaths, with gravel or any clean material, and formed with a transverse inclination away from the reservoir.

The outside of the puddle wall should be protected from rats by a facing of concrete, or else the earth may be mixed with lime in the proportion of about 10 to 1, and watered for a yard in width outside the puddle wall, and this should be extended along the surface of the ground for some part, if not the whole, of the seat of the outer part of the embankment.

In the case of a high embankment, as of a storage reservoir, this protection is not necessary, because the distance from the outside to the puddle wall is too great to be burrowed, but in small service reservoirs the puddle should be protected.

In many situations where it would be proper to construct a service reservoir, it would be difficult to procure clay for puddle within any short distance, and it might, at the same time, be easy to procure materials for concrete. In that case the reservoir might be made watertight, without puddle, by the use of concrete made specially with that view, that is, with plenty of filling, both sand and lime or sand and cement; but if good blue lias lime can be procured at a less price than cement, it will be preferable to use lime, the quantity being increased in the ratio that the price of cement bears to that of lime; and the thickness of concrete may be less than that required of puddle; and, reckoning the thickness of puddle and its covering together, this is a considerable reduction of the thickness of materials required for the bottom of the reservoir, and saves excavation to the extent of about half the difference of the cubic contents.

If, however, there is the slightest reason to fear a movement of the ground, however slight or partial in its extent after the reservoir has been made and filled, puddle will be preferable to concrete as being less rigid. At the same time the longitudinal tenacity of concrete may be a favourable property in the opposite respect. The earth would be wheeled out, or carted out, and put into the banks in thin layers not more than 6in. in thickness, and consolidated by hard ramming, and, if sand, watered at the same time.

A sand bank made in wet weather is very much superior to one made in dry weather.

If the banks be wholly of sand the puddle walls should be of greater thickness than is otherwise necessary, unless they be faced with clayey material, or with limed earth to prevent the moisture of the puddle being absorbed by the banks.

Service reservoirs, larger and deeper than the one of this example, have been made on ground containing nothing but sand, and with the strength-dimensions here stated, stand firm and watertight; but ground containing sand, clay, and stone shale mixed is preferable, if the clay be not in excess. Instead of the expensive brick paving of the inside of the reservoir, a facing of Portland cement on concrete would be almost equally good, and considerably less expensive.

A particularly good fence is necessary round the foot of the slopes. Post and rail is good enough for a storage reservoir, but not for this, and the fence should be far enough from the foot of the slope to allow a cart-way all round, and a piece of land should be inclosed near the outlet of the reservoir for stores and other purposes.

SECTION XXIV.

Distribution of Water.

The main from the service reservoir to the town has a great declivity if taken as a whole, but for a short length after leaving the reservoir it sometimes lies above the hydraulic incline, and will not supply so much water to the remainder of the pipe as it is capable of conveying. Seeing this, it has in some instances been sought to obviate it by making the diameter of the pipe greater at its upper end than further on. This seems right in principle where the main does lie in this position, but if it descend to or below the hydraulic incline immediately after leaving the reservoir, it is not so; the pipe must then be of the same diameter throughout. In either case the mouth of the pipe should be enlarged so that it will admit the full quantity due to the carrying capacity of the main.

The trunk main should be continued through the town of the full size, so that future branch mains may be derived from it at any point that may hereafter be found desirable.

Branch mains are commanded by sluice valves at their junctions with the trunk main, so that the town becomes divided into districts, and the water may thus be shut off from any one of them without affecting the supply to other parts of the town. Waterworks valves are generally of the screw kind. A watertight cast-iron box contains a gun-metal nut which is attached to the door, and through which works a gun-metal (but sometimes wrought-iron) screw, with square thread, by which the door is raised, the door being faced with a gun-metal ring on the side opposed to the pressure, which slides on another gun-metal ring attached to the cast-iron body of the cock. The head of the screw should have a false head of iron to take the wear of the key.

The very various rate of flow in town pipes caused by the unequal quantities drawn for domestic use at the various hours of the day renders it a matter of difficulty to calculate exactly the proper sizes of the pipes. Mr. Marten* made some experiments at Wolverhampton under a constant supply, and found that in round numbers three-fourths of the whole daily supply was used between the hours of 8 A.M. and 8 P.M.; that the hourly quantity used during the day increased from 4 per cent. of the whole at 6 A.M. to 8 per cent. at noon; then only about 6 per cent. per hour is used during the next two hours, but at the end of that time the increase begins again, and reaches 8 per cent. per hour at 4 o'clock, after which it declines gradually to 3 per cent. at 9 at night; and during the nine hours of the night, from 9 P.M. to 6 A.M., the supply is only 14 per cent. of the whole, or at the rate of $1\frac{1}{2}$ per cent. per hour.

Mr. Hawksley stated, in his evidence before the Committee of the House of Commons upon the Salford Improvement Bill, that during the week ending February 2, 1862, at 9 o'clock in the morning the pressure at the entrance of the main into Salford was equal to a head of water of 173ft., which was reduced to 169ft. at 10 o'clock, rising again to 170ft. at 11, then reduced during the hour before noon to 168ft., and, the people being at dinner from 12 to 1, and no water being used at the mills and other works, the pressure rose during that hour to 172ft., falling again to 158ft. at 2 o'clock, when the people had returned from dinner, rising again to 160ft. at 3 o'clock; and at 4 o'clock a change is made by connecting the main with a reservoir at a lower level than that which supplies the day pressure.

The head from this lower reservoir remains on during

* Henry J. Marten, Esq., M. Inst. C.E.

the night and until 6 or 7 o'clock in the morning, when the pressure from the higher reservoir is again put on. At 5 o'clock P.M., the pressure from the lower reservoir was equal to a head of 75ft., and it rose again during the evening, as people ceased to take water, to midnight, when it was 98ft., then, a few people being still up and drawing water, it rose still further after that up to 102ft. at between 2 and 3 o'clock, where it remained until 5 A.M., after which, a few people getting up at that hour and drawing water, it was 100ft.; and at 6 or 7 o'clock the pressure from the higher reservoir was again put on.

These observations show the hours of the day at which the maximum and minimum quantities of water were being used. The greater the quantity being drawn off the more the pressure is lowered in the main.

It has also been stated that in general the draught of water in a town between 8 o'clock in the morning and noon is usually from 2 to $2\frac{1}{2}$ times the average of the whole day. In the night it is not more than $\frac{1}{4}$ of the average quantity, except there are cisterns to be filled up during the night, and then it is not so low as $\frac{1}{4}$.

The quantity varies at different times of the year; the greatest quantity is used in the month of August, except in London, where the greatest quantity is used in July.

The appendages of town pipes are sluice-valves, cleansing cocks, hydrants, meters, lead or iron house-service pipes, with their ferrules of attachment to the street pipes, stopcocks, bib cocks, and waste preventers.

The old conical plug cock was difficult to make water-tight, and is now superseded by the screw-down cock, with a leather seating. There may be a second or upward-action button valve which rises with the flow of water to a seat under the seat of the upper valve and allows the upper valve to be taken out for repairs while the pipe is under pressure.

The old system of obtaining water from the street pipes for the extinction of fires was by a simple hole in a projection cast on the pipe, which was closed by a wooden plug; hence the common name of these appendages—fire-plugs, and they are yet in use in London, but in general they are superseded by the hydrant, which has the opening closed by a ball which is floated up to an india-rubber seat by the water, and by its pressure the pipe is kept watertight, and when it is required to obtain water from the pipes this ball is pressed down by a spindle (carrying a cup at its lower end) which passes through the centre of a tube having means of attachment to the pipe at the bottom and nozzle outlets at the top to which hose pipes are attached, the water escaping past the ball when it is pressed down and rising through the tube. The same appendage is also the most convenient for filling carts for road-watering.

Recent observations on the quantity of water required for road-watering show that 220 gallons of water are expended on 1,088 square yards of macadamised road each time of watering; that is to say, that quantity watered 272 lineal yards of road 4 yards wide, or half the width of an average street, which is at the rate of 200 gallons per 1,000 square yards watered, or where the roads are watered twice a day, as they were in the town where these experiments were made, 400 gallons per day per 1,000 square yards watered. The quantity used by each cart was estimated thus—hydrants being placed all along the roads at the distance of 272 yards apart, so that there is no loss of time in dragging full carts over any part of the roads already watered.

Take the time at which a cart begins to be filled. It takes four minutes to fill it, disconnect the hose, and start. It then takes four minutes to travel and expend its load, by which time it will have arrived at the next hydrant, where it takes up another load and travels on as before, thus keeping on in procession, and not uselessly travelling over the ground. In this way a cart holding 220 gallons will water $2\frac{1}{2}$ miles of road twice a day over the full width.

The quantity stated to be used in London is 14,000 gallons per mile of road watered twice a day.

Water meters are of two kinds; the one those which measure positively and the other those which measure inferentially. The positive water meters measure by the cylinderful, the piston, which is driven by the pressure of water through the fixed length of the cylinder, registering each stroke as being the quantity of water displaced by its motion.

The inferential meter is on the principle of a Barker's mill—that is to say, the revolution of a spindle is caused by the unbalanced pressure of water within radial tubes which have holes on one side only, the pressure due to the area of the hole being transferred to the opposite side of the tube, causing it to recede from the direction of the hole with the velocity due to the head of water and area of orifice.

To bring this principle into a practicable shape the water passes down a funnel and outwards through curved arms, the areas of whose outlets are proportioned to the quantity of water to be registered at each revolution, and inasmuch as the number of revolutions would increase with the pressure of water, and therefore register more water under great than under small pressures, vanes are attached which, by meeting with a greater resistance under a rapid than under a slow velocity, retard the motion in the same ratio as the greater head of water tends to increase it, thus registering the same quantity under all pressures.

By these means and by minute attention to the perfection of manufacture of every part of the machine great accuracy is attained; the meters are indeed guaranteed to register accurately within 5 per cent.

A water-waste preventer is a most necessary appliance, and consists of a cistern in two compartments—a larger one into which the water is received, and from which it is shut off when full by the rising of a floating ball at the end of a lever, and a smaller one into which water is admitted from the larger compartment, the two plugs closing the outlets being so connected that both outlets cannot be opened at the same time, but that when one is opened the other is by the same motion closed, so that a continuous flow of water through the cistern cannot take place.

The service pipes laid into houses are not usually connected with the mains, but with sub-mains or service-mains supplied from the mains by junctions at intervals, separate districts being thus formed, the service-mains of each district having a stopcock upon it near the junction with the main; so that when it is desired for any purpose to shut off the water from a district it can be done without stopping the supply to other parts of the town. This is one of the oldest devices in the water-supply of towns, but in adopting the system formerly it was more a matter of necessity than it is now in many cases. The intermittent system of supply was the general one, formerly, under which the water is delivered into cisterns on the house-premises, being turned on to a district for a certain number of hours in the day and turned off at night, to save leakage from the street pipes as well as from the house-service pipes. But this system of sub-mains, or service-mains, each controlled by a stopcock, is a convenient one, whether the system of supply be the intermittent system, as in London, or the constant system, as in some other towns. It also serves the useful purpose of enabling the quantity of water supplied to each district to be measured. The method by which this has mostly been done is to insert in the service-main behind the stopcock a short pipe carrying a hydrant, and another short pipe in front of it, also carrying a hydrant, a standpipe being attached, temporarily, to each hydrant, and their nozzles connected by a pipe with a meter upon it, through which the water is turned by shutting down the stopcock, the meter and the two standpipes being removed when the measurement has been made, and the stopcock opened for the passage of water along the service-main as usual.

The indicator dials of the meters, whether they have rotary or linear motion, show moderately small quantities of water passed through them, and by observing the differences of the times of observation and making those differences sufficiently small, an accurate account of the flow for any length of time—say 24 hours—can be produced. When it has been desired to change the system of supply from an intermittent to a constant supply it has always been an anxious question what effect this would have upon the total quantity at command, and this has been ascertained by making trials in the separate districts, before disturbing the whole arrangements, by placing a meter upon the service main and comparing the population supplied with the quantity registered in 24 hours, and continuing the observations for a whole week, by which the engineer could ascertain whether it would be safe to advise that the change of system should be made.

When such observations have been made it has usually been found that a large quantity of water was passing through the meter even during the night time when it was known that water was not being used, and the quantity then passing into the district could only be attributed to leakage, either from the service pipes of the houses and their fittings or from the service mains in the streets, but from which of them the leakage proceeded could not be discovered by the readings of the meter, which gave only an integral number of gallons in a given time, and which time would be considered reasonably short if taken every quarter of an hour; and by a system of night inspection and a strict ordering of the prevention of waste by leaving taps open or cisterns overflowing, the sources of leakage could often be traced to defects in the street pipes belonging to the water authority; and before a constant supply could be attempted these defects were first repaired, and sometimes the pipes were renewed altogether, there remaining then only the waste produced by defects in the house-service pipes and their fittings; but as householders object to frequent inspections of premises by the officers of the water authority, it still remained very difficult to ascertain upon which particular premises the leakages occurred. This difficulty has been in a great measure obviated by the means taken first at Liverpool.

Mr. G. F. Deacon, the water engineer of Liverpool, invented a new kind of meter for the purpose of indicating the waste due to leakage, by which the quantity passing through the meter is recorded continuously, instant by instant, describing a diagram of the movement of the water through it. The essential part of Mr. Deacon's meter consists of a conical tube, truncated, with a disc of the same diameter as the smaller end of the tube, moving rigidly in its axis, the water entering at the smaller end of the tube and leaving at the larger end. The conical tube being vertical, its small end uppermost, the weight of the disc is counterbalanced by a weight hung from a pulley above it, the connection being made by a fine wire passing through a pinhole in the top of the meter. When no water is passing the disc fills the small end of the tube. On the water being turned through the meter the disc is driven away from the small end of the tube by the pressure against it, leaving an annular space between its edge and the tube round it, through which the water passes. A pencil on the wire, connecting the disc with the counter-weight, marks the rise and fall of the disc upon a sheet of paper wound round a cylinder, which is made to revolve by clockwork. The paper is wide enough to show the greatest average quantity per hour at any instant, and long enough to record the flow of water into the district during 24 hours. By reason of the sensitiveness of the disc, suspended in a stream of water of very unsteady flow, as small a quantity as that issuing from one tap in a district containing perhaps five hundred, or more, is shown on the diagram; indeed even a smaller quantity than this is said to be distinguishable sometimes, so small even as that of a strong dropping of water from one tap. Such small variations of the movement, however, might indicate variations of pressure on the disc produced by the alternate compression and release of the confined air in the service main and pipes. But this is a trivial matter in comparison with the important service rendered by this meter to all water authorities in ascertaining the sources of waste.

Rivers.

For waterworks purposes a river may be considered to be a natural conduit which conveys water from a reservoir, but that the reservoir from which the conduit proceeds is spread out over an extended and shallow area, or else is subterranean; and that the discharge is not artificially controlled. In this case we have less to do with immediate drainage area than with the actual volume of the river in the driest seasons. This dry-weather flow is determined by frequent actual gaugings of the river after long droughts. So that, except for scientific purposes of investigation, the drainage area and rainfall do not immediately enter into consideration.

But besides finding that the dry-weather flow of a river is sufficient to allow water to be taken from it for the supply of a town without injury to interests below the point of abstraction, or with such small injury as may come within the possibility of direct money compensation, there arises the question of purity of the water; whether, allowing that no river in its lower or middle courses can be absolutely pure, it is sufficiently so for a town supply. It is evident that many rivers have been so much polluted by manufacturing refuse and town sewage during the last twenty years or more that they are quite unfit for the purpose, but, besides these, there are those which have long been considered doubtful; and the state of the Thames and Lea, from which London is chiefly supplied, has of late years given rise to much discussion in this respect; and it is to be presumed that other rivers from which water is supplied to towns will receive a similar attention. The question is far too great a one to be treated of here exhaustively, and all that need be said is that the Royal Commission on Water Supply reported in 1868 that, notwithstanding the impurity the Thames received in its course before arriving at Hampton and Kingston, where the supply of five out of the eight water companies was taken, it was sufficiently pure to be supplied to the metropolis. The quantity is abundant for the requirements of London, and the companies are authorised to take about one-fourth of the dry-weather flow of the river, that being about 400,000,000 gallons per day, before it reaches the waterworks, and that will be, when the companies have taken the 100,000,000 gallons per day they are entitled to take, about 300,000,000 gallons per day still; and there are no interests below them injuriously affected by the abstraction of that amount of water except that of the Thames Conservators in respect of the diminution of the freshwater flow as far as that may affect the ebb of the tide from Teddington to, say, Putney; and these interests have been arranged with the companies by each of them paying to the Conservators £1000 a year, conditionally on their taking measures to prevent the flow of town sewage into the river. The average volume of the river, however, is three times the quantity here set down.

But although the Thames, from its peculiarities, is able to furnish a good supply of water to London, in the opinion of the Royal Commission, there are other rivers in the country from which towns are supplied which will probably, on a like investigation, not prove to be so suitable. Indeed rivers, which for many years have been the sources of water supply, have been given up from time to time, and purer sources resorted to.

Water is pumped from a river into a service reservoir through cast-iron pipes, mostly by steam-engines; in a few cases by water-wheels: but scarcely any river water is pure enough without filtering, and it is usual to filter it before it is pumped up to the reservoir. It is seldom that depositing reservoirs and filter-beds can be constructed on the river side so that the water may flow into them directly from the river, and in most cases the water is pumped into depositing reservoirs and filter-beds formed on or near the banks of the river, and in these cases the engines have a double set of pumps, one to lift the water from the river into the depositing reservoir, from which it flows back to the pumps through the filter-beds, and another set of pumps to raise the water to the service reservoir.

WELLS.

The chalk and New Red Sandstone formations absorb large quantities of rainfall, and in those situations where there is no free escape for the water in the form of springs, as is the case where the strata lie in a basin-like form, or where faults hold up the water, shafts sunk into these strata often yield large quantities of water. The Kent Waterworks Company were said some years ago to be pumping 6,000,000 gallons per day, and that they could, if required, obtain 10,000,000 from the same site.

At Liverpool there were seven wells sunk in the New Red Sandstone, from which 4,000,000 gallons per day were pumped, while 2,000,000 were being pumped from private wells, making 6,000,000 gallons per day, all within an area of six square miles. The Corporation abandoned three of these wells, continuing four in use and sinking a new one, and from these five wells 38,000,000 gallons per week were pumped, according to the evidence of the late Mr. Duncan in 1867. This would average a little over 1,000,000 gallons per day from one well, but there is nothing more certain than that wells sunk in the New Red Sandstone vary greatly in the quantity of water they yield; as, for instance, while the average of these five wells was not much more than 1,000,000 gallons a day, one of them, that, namely at Green Lane, yielded 3,000,000 gallons a day at the same time. The supply of water to be procured permanently from wells is problematical whatever the ground they be sunk in. The quantity which can be permanently derived from wells is more uncertain than that from either rivers or gathering grounds of moorland at considerable heights above the sea.

SECTION XXV.

Pumping Mains and Engines.

THERE is not much difference in the conditions of a pumping main through which water is supplied into a service reservoir, and those of a conduit pipe, except the shocks and variations of pressure to which it is subjected unless good arrangements of air vessels be adopted at the pumping station, and even with these it requires to be of rather greater proportionate strength than a conduit pipe; and when it is considered that in general a pumping main passes through a more numerously-inhabited district than a conduit pipe usually does, and therefore that, in case of a burst, so much more damage would be done, there are obvious reasons for making it considerably stronger.

When branches are derived from the pumping main, it should still be continued of the full diameter up to the service reservoir, because the flow in the branches is liable to checks, which would throw strains on the pumping machinery if there were not a free way for the water at all times, the consequence of a check in the branches being, under these circumstances, only that more water would then be delivered into the reservoir.

Where water was pumped into a main from which branches and service pipes were derived in its course, and it was proportionately reduced in size after giving off these branches, it was sometimes tried to equalise the strain on the pumping machinery caused by these checks to the flow by pumping the water over a stand pipe of the greatest height required to supply any part of the district; and this, of course, did equalise the strain, but it was not an economical method of working, for all the water had to be pumped to the maximum height, while the greater part perhaps required no more than half the elevation to effect a due delivery; and then, besides this bad economy, the water would often stand in the down leg of the stand-pipe at many feet below the summit, and the body of water falling down freely on to its surface caused concussions that were felt throughout the whole system of town piping. Stand pipes are not so much in use now, and it is preferred to pump direct into the mains, with the provision that its size be continued full to the end at an elevated spot where a reservoir can be constructed. That pumping directly into a system of town piping not provided with a head reservoir is injurious to the plant was shown at the Liverpool works, where, according to the report already named, the fluctuation of the load on the engine at one of the pumping stations varied fifty times in a minute from 104ft. to 184ft., when the engine was making 25 strokes per minute. The machinery seemed as if it were being knocked to pieces.

The two forms of engine which have most often been employed for pumping water are the single-acting Cornish engine, with a large steam cylinder at one end of a beam, and a heavy plunger-pole at the other, the characteristic movement being a long stroke quickly made, and then a pause of greater or less duration before another stroke is made, according to the number of strokes per minute required for present work, the frequency of the strokes being under the easy control of the attendant by the means provided; and, secondly, the double-acting, double-cylinder or compound-cylinder beam engine with a crank shaft carrying a heavy fly-wheel.

In all pumping engines, of either form, a heavy mass must be put in motion in order to overcome and equalise as far as possible the variations of the load caused by the different pressures from time to time in the pumping main, except in cases where the water is pumped into a reservoir through a main from which no branch is derived, and even in these cases, unless the engine be provided with differential valve gear, so as to stop the motion at once automatically, as with Mr. Davey's gear for that purpose, a heavy mass put in motion is still required in order to guard against breakage of the engine, in case of a burst of the main. For pumping water it has been said that the Cornish form of engine is the best, but it would rather perhaps be a question first, what are the particular circumstances under which the engine is to work.

Formerly the water was in many cases pumped over a standpipe, high enough to command the highest parts of the low-level districts to be supplied; the smaller quantity of water required for exceptionally high parts of a town being pumped into and supplied from a high-service reservoir, the water being turned off from the standpipe during the time of pumping into the reservoir. But the districts into which a town is divided for the purposes of water-supply do not draw from the main the same quantity of water at all times, and when they are drawing most quickly, the level of the water in the descending legs of the standpipe is lowered very much, and the water falls from the top of the standpipe with a momentum which is enough to burst a pipe sometimes. If, for instance, the standpipe be 200ft. above ground, and the surface of the water in the outer legs be 150ft. only, there will be a drop of 50ft., and the water will attain a velocity of more than 50ft. per second, causing dangerous concussions in the pipes.

There is another objection to standpipes; the engine must pump all the water to the full height, although so great a head may not at all times be required. In the Cornish engine employed in pumping water from a mine the load is the same at every stroke, and the standpipe was introduced to resemble that condition of working, but a balance of practical disadvantages has been against it, after being many years in use, and the air vessels and a better arrangement of pump valves are made to take their place in the regulation of the load on the engine, whether it be of one or the other type here mentioned. The kind of engine first employed was the single-acting beam engine with crank shaft and fly-wheel, which was a safe engine for the purpose, but not economical in the working expenses. As the Cornish engine was highly economical for pumping water from mines, it was adopted also for pumping water for towns where the circumstances were at all similar, and sometimes perhaps where they were not so; and when, about the year 1850, the late Mr. James Simpson, the engineer of the Lambeth Water Works, desired to pump water through a 30-in. main from Thames Ditton to Brixton Hill, a distance of more than 10 miles, he requested Mr. Pole, F.R.S., M.Inst.C.E., and the late Mr. David Thomson, to undertake an investigation into the merits of different forms of engine, with the view to the adoption of the one most suitable to the conditions under which the water had to be pumped.

At that time, as may be seen in the Minutes of Proceedings of the Institution of Civil Engineers, vol. xxiii., the Cornish single-cylinder expansive engine, which had been introduced into London by Mr. Wicksteed, had been adopted by several waterworks engineers, and had justified the reputation of Cornish engines for economy; but as none of these worked under the same or even similar conditions to those in this case, Mr. Simpson determined to have the whole subject investigated, with a view to ascertain whether the other form of expansive engine, the compound cylinder, would not prove more applicable; but the investigation proved that in the double-cylinder or compound cylinder engines which were then in use in various parts of the country, the expansion of the steam had not been carried to a sufficient extent to produce great economy, nor had it been arranged in the best manner to attain equality of motion, and the arrangements of valves and passages were generally so defective as to cause great loss of power, and waste of fuel; but an attentive study of the principles of this form of engine led Dr. Pole and Mr. Thomson to the conclusion that, with a well-considered design, carefully carried out into practice, the compound-cylinder arrangement promised to offer a more beneficial application of the principle of expansion to rotative engines than could be attained by the single-cylinder form; and when the Lambeth Water Works extension scheme was carried into effect, large engines of 600 H.P. were carefully designed, and as carefully executed, and the result fully justified the expectations.

Trials made at the Lambeth Water Company's Works in February 1884, by Mr. Taylor,* the present Engineer to the Company, of engines constructed on a similar principle, were as follow: The average head of water on the pumps was 187·65ft.; the pressure in the large boilers was 61lb. per square inch, and in the jacket-boiler 101lb.; the barometer stood at 30in., and the vacuum at 28in.; the trial extended over the time during which 12,956 revolutions were made, and the coal consumed during the trial was 2,090lb. in the large boilers, and 298lb. in the jacket-boiler, being together 2,388lb. These trials and results are summarised in a statement by the makers of the engines, Messrs. Simpson and Company, of London, as follow: 27 revolutions per minute; actual horse-power, measured by the water lifted, 167·6; coal consumed per actual horse-power per hour, 1·784lb.; duty per 112lb. of coal, 124,270,000 foot pounds; average indicated horse-power 193·6; coal consumed per indicated horse-power, 1·54lb.; quantity of water pumped during the trial, calculated from the full capacity of the pumps, 1,412,204 gallons.

These results agree closely with others obtained in a trial during 24 hours in October 1881 by Mr. E. A. Cowper, M.Inst.C.E., with two engines of the same construction at the same works, made by the same makers, which were as follow: The feed to the boilers per indicated horse-power

* John Taylor, Esq., M.Inst.C.E.

per hour was 13·397lb.; the coal consumed per indicated horse-power per hour, including that for the jackets, was 1·6049lb.; the amount of heat, including that given to the jackets of the cylinders, was 10,100 thermal units per lb. of coal consumed.

In a 48 hours' trial, made in February 1883 with similar engines at the Hammersmith Pumping Station of the West Middlesex Water Works, by Mr. Thomas Hack, M.Inst.C.E., the Company's Engineer, the value of the first condition mentioned above was 14·67; of the second, 1·54; and of the third, 11,675.

The trials made by Mr. Taylor, as given first above, may be compared with those of Mr. Hack and Mr. Cowper, on the same basis, as follow, viz., the feed of the boilers 13·67lb. per indicated horse-power per hour; the coal consumed was 1·54lb., and the heat produced per lb. of coal consumed was 11,682 thermal units in the large boilers, and 7,500 in the jacket-boiler. It may be said perhaps that the high duty of 124 millions of pounds raised 1ft. high by the consumption of 1cwt. of coal is partly due to the high quality of the Nixon's navigation Welsh coal used, but the makers of these engines say that in all these trials the steam condensed in the jackets returned to the boilers, and the feed, which was most carefully measured in tanks, is the weight of steam passing through the cylinders; and that this method of testing the efficiency of engines enables comparisons to be made of their relative economy without regard to the quality of the coal used.

A duty of 88 million foot pounds with the consumption of 1cwt. of the best coal is perhaps as much as can be done with the Cornish form of engine, and this is equivalent to a consumption of 2·5lb. of coal per hour per indicated horse-power, so that the working economy of the compound cylinder rotative beam engine seems to be the greater of the two.

SECTION XXVI.

FLOODS.

To prevent floods, the first thing to be done of a practical nature is to ascertain the quantity of water which the river will carry off when in its normal condition. The normal condition of a river must be taken to be not its original condition, and not, in most cases, its present condition, with all its obstructions to the full flow of water, but such a condition as that would be which would allow the continuance of weirs established for useful purposes in suitable situations, and the construction of which is such as not to obstruct the passage of flood waters more than is necessary to the normal conditions established on any part of any river. All other obstructions would be removed. They consist chiefly of accumulations in the bed, arising from continued neglect of the regimen of the river. Every summer the banks are trodden down by cattle, every autumn there is a general decay of vegetation both on the banks and in the bed, and every year the heavy rains carry earth into the river from the high ground. When the river is unobstructed by decayed vegetation, the earth brought into it during heavy rains is carried forward with the water and deposited in the slack water of bends and in the pools of dams and weirs, a portion being further carried into the estuary of the river; and all these movements and deposits are consistent with and form part of the proper regimen of every river; but when weeds are left in the bed in winter where they grew in summer; when the injury to the banks done during summer by cattle treading them down is not repaired in the autumn; these catch the earth brought down in heavy rains, and an agglomeration of decayed vegetable matter and earth is formed, which floods do not carry down the river, but move about to short distances, forming shoals. But obstructions in the bed of a river are not in all cases due to neglect; it sometimes is the case that the ground in which the river has formed its course is of too hard a character to be scoured to the necessary depth by the action of the water alone, and this is the case not only here and there, where a ledge of rocks runs across the bed of a river, but for long distances, or where it consists of hard boulder clay; and here the bed of the river would be lowered by excavation.

The prevention of floods means practically the prevention of the injurious overflow of rivers, and this is effected by various means, the chief one being the lowering of the permanent water-level by deepening the bed and making such defences in certain places along the banks as to enable the water itself thereafter to maintain the depth by scour in flood times.

Another means is the storage and regulation of the excess of water over and above the carrying capacity of the river when its bed has been duly lowered, and means taken to prevent accumulation of the detritus in other places than those provided for it; but this aspect of the question, in connection with river improvements, has not always been considered in its due relation to the possibilities of things. The storage required is not of the same nature as that required for the supply of towns, except subsidiarily, and for a comparatively small portion of the water to be dealt with; it is rather the regulation of the flow of the excessive rainfalls than the absolute conservation which is, in this case, the object of "storage," and this takes it out of the category of those impossibilities which have sometimes been said to exist in this matter. The first thing, then, to be ascertained is the quantity of water the river will carry off in every part when its regimen has been restored and amended through those lands which are liable to floods; and then to regard the storage and regulation of the flow of the remainder.

It has been said that these are separate questions; but they cannot be so, and unless they be considered together, in the first instance, and before anything is done in the improvement of the river itself, it may well be apprehended that either too much or too little will be done in that respect. The apprehension lies not so much against the engineer as against the politician, who may, perhaps, limit his view of the water question, in respect of floods, to the improvement of rivers only, regardless of the concurrent question of the regulation of that further quantity of water which the rivers cannot carry off as fast as it falls. If he do the one thing only, he will have to make an undue interference with existing water rights in the river, and, besides that, he may damage the adjoining land in dry times more than would be compensated by their relief from floods. By this, too much would be done; and if, on the other hand, the water rights be duly respected, and at the same time the required moisture in the adjoining lands be maintained in dry weather, too little would be done, for floods would still be frequent; and the water question is not one in which a given expenditure of money confers a benefit strictly in proportion to its amount.

In considering the question as a whole, the circumstances of the case should be considered downwards from the summit of every hill, and a comparison should be made between the original and natural state of things and their state to-day under the requirements of the inhabitants. There is plenty of evidence that the higher portions of the land we inhabit have been denuded of their finer particles and fragments by the force of water, falling in the first place as rain, which have been, and continue to be, carried down towards the sea. Whether this has always been so or not, it is certainly true of the present time and circumstances, and may without doubt be said to have been true of the circumstances of river basins before men became the proprietors of the land. It is between that time and the present that it is desirable to institute a comparison, in order to see whether the owners of the higher lands should, in justice and reason, contribute to the relief of the lower lands from floods. The denial of the upland proprietor that he contributes in any way to the injury of the lower lands, and that he only transmits what falls upon his land by no act of his own, holds good in respect of the water, but of the water only. If the upland proprietor allows the soil and fragments of his land to be discharged into the river, he does not exercise the power he possesses to maintain his property without injury to his neighbour. It is not to the point, and does not satisfy the sense of justice, to say that he cannot prevent the soil being washed away by the rains in the natural way. He claims to exercise a right over the soil, and the sole right, to the exclusion of whomsoever, and it could easily be shown, as against him, that if he will he can keep the soil and *débris* of his land upon his own premises. That he will not go to the expense of doing this is reasonable enough, but it spoils his argument that he does not contribute to the injury of his neighbour.

To revert to the original order of things, the natural conditions are that the *débris* of the higher lands is washed down into the river by the rain, but also under the natural conditions of the river it was carried towards the sea, much of it being spread out and deposited along the valleys by the heavier rains, raising the land into the present meadows. Many proofs of this are found in the excavations which have been made for sewers, bridge foundations, and other works in valleys, where old roads and implements have been found at a depth of several feet below the present surface—as much as 5ft., certainly. This, of course, amounts, even in one river-basin, to a vast quantity of earth deposited in the valley, but probably much more has been carried down into the sea by the river, the heavier rains causing such a scour along the bed as to maintain a sufficient channel.

The extent to which a weir influences the height of the water in the river above it—the amplitude of the backwater—can be found approximately for any weir, the circumstances of which are known, together with those of the river. The accompanying sketch is a section of such

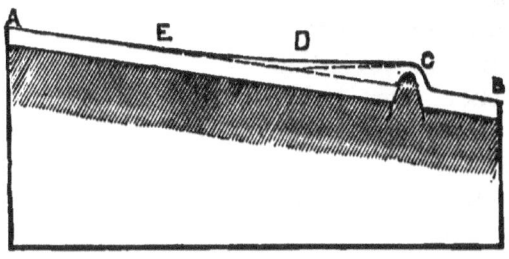

Fig. 55.

part of the river above a weir as is influenced by the backwater, the vertical scale being greatly disproportionate to the horizontal.

If A B represent the surface of the river in its original state, and a weir or dam be erected in the position shown, the water will pass over it at a height C due to the quantity and the length of the weir. If from that height C a line be drawn horizontally backwards until it cuts the surface of the original flow at D, as represented by the dotted line C D, that distance, from C to D, will be about half the distance to which the surface of the water in a regular channel will be influenced by the erection of the weir; that is to say, it will be so influenced as far up as the point E, twice the distance C D. The authorities give the distance C E as being from 1·5 to 1·9 times the distance C D; but if twice that distance be taken, it will allow a margin for error, and will give the utmost limit of the backwater. From that point upwards to the next weir the river remains uninfluenced by the weir below; but it is only in rivers with considerable fall, or where there are but few weirs, or when the weir raises the height of the water but little above its original level, that there is any great length of river thus uninfluenced, in the ordinary state of the flow of the river; but the influence of the weir becomes less and less as the quantity of water coming down the river increases, until, in a very high flood, when the weir becomes drowned, its influence in checking the flood is but little.

The late Mr. E. Leader Williams gave evidence before the Select Committee of the House of Lords on Conservancy Boards in 1877, that when the Severn Navigation Commissioners were before Parliament for authority to construct these weirs the question was fought over for nearly six weeks, and the result was that he arrived at a form of weir which not only pens up a certain quantity for the purposes of navigation, but facilitates the passing off of the flood water as well. It is a solid weir, with a level crest from end to end, and without sluices or means of passing off the water other than over its top. The back of the weir is rounded off to a parabolic curve, instead of being, as most weirs of the kind are, flat on the top. This improved form gives greater velocity to the water passing over the weir, and reduces its height for any given quantity passing over. This, however, is not the curious point to which allusion is made above; it is that before a weir is formed, the flood-water runs over a rough bed, there being at the first of the flood but little water in the river—but meaning, presumably, only in the case of the Severn, or similar rivers—and, therefore, meeting with great obstruction to its flow from that roughness; whereas, when the weir is erected, it forms a pool of comparatively still water above it, through which the flood passes more readily than it did before; indeed, directly the first of the flood touches the upper end of the pool, it sets in motion the whole mass of water, and it is discharged more rapidly than it would be if it came into an empty channel.

This was supported by Mr. J. H. Taunton, the engineer

of the Thames and Severn Canal Navigation. It has been proved to have been practically so on the Severn, and that is much to the point; but it would not be justifiable to consider this proposition as having reference to rivers in general. In order to still further reduce the height of the water due to any given quantity passing over the weir, the length is greatly increased by constructing it aslant of the river course, making it, for a river 150ft. wide, 500ft. long, or thereabouts; and besides the effect which the increased length has in reducing the height of the water passing over it, the form in which the back of the weir is constructed, being that of a parabolic curve, has another effect in the same direction, for, compared with a weir which has a flat crest, this form of weir causes an increased velocity of the water in the ratio of 5 to 7. This is so when there is a free fall over the weir, but in high floods the boat traffic passes over these Severn weirs instead of through the locks, the weirs being submerged and the water-surface nearly level.

A great flood in a river, in its upper part, where it is no bigger than a brook, albeit a large one, is from 12 to 24 cubic feet of water per minute per acre of the few thousands of acres of ground above the point of observation, and from which the water flows; and sometimes it is as much as 30 cubic feet per minute per acre from relatively small areas. Excessive rainfalls are but partial in their extent over large river basins. These quantities are known because it is here, in such situations, on the high ground within some few miles of the extreme watershed, and within the few thousand acres of the upper part of a river basin, that reservoirs are made for the storage of water for the supply of towns and for compensation to mills, and, formerly, for feeding the canals in dry seasons; and these reservoirs have afforded opportunities for measuring the flow of large quantities of water proceeding from known areas of ground. When a reservoir is full, the flood may be measured by the quantity going over the waste weir, and that passing at the same time through the outlet pipes, and when the water in the reservoir is considerably below the top-water level on the commencement of a heavy rainfall, and the height is known at which it stands in the reservoir, the quantity of water flowing off the ground is measured by the capacity of that portion of the reservoir filled up in the time observed, and as it is usually known what quantity of water is contained in each foot in height of the water-level, the volume of the flood is at once ascertained.

If the volume of flood-water per minute be measured at various points on the course of a river between the watershed and the outfall, it will be found to be much greater in the upper part of the basin than in the lower, per acre of the drainage ground down to the point of observation.

Within those areas appropriated to the filling of impounding reservoirs with flood-water, within, that is to say, an area of ten thousand acres, the volume of a flood may be from 24 or 30 to 12 or 15 cubic feet per minute per acre; but at the outfall of a river a great flood is not more than 1 or 2 cubic feet per minute per acre, as may be seen from the few instances following.

The area of the Thames basin above the tideway is 3,676 square miles, which is 2,352,640 acres. Mr. Hawksley estimated the maximum flood in the Thames, in his evidence before the Water Supply Commission, at from 20,000,000,000 gallons in a day to 25,000,000,000, which would be from 2,222,222 to 2,777,777 cubic feet per minute: and if we take the mean of these two quantities it would be 2,500,000 cubic feet per minute, or 1·06 cubic feet per minute per acre; or if we take the extreme quantity it would be 1·17 cubic feet per minute per acre. This is at a distance from the source of about 100 miles, not following the windings of the river, but a general line down the valley.

At the Albert Bridge, near Windsor, Professor W. C. Unwin gauged, in 1875, one of the highest floods which had occurred there for many years. The quantity was

14,094 cubic feet per second (*vide* Minutes of Proceedings Inst. C.E., vol. xlix.), or 845,600 cubic feet per minute.

Mr. Unwin does not state the drainage area to that point of the river, but tracing upon the map of the catchment basins, to which allusion was made in a former section, a line from that point to the watershed on each side of the river, dividing off the lower portion of the area, there appears to be a watershed area down to that point of 1,700,000 acres, and according to this the quantity would be ·5 cubic foot per minute per acre, very nearly. The distance down the valley is about 80 miles.

On the Severn, at Worcester, the late Mr. E. Leader Williams, the engineer to the Severn Navigation Commissioners for many 'years, gauged the highest flood he had known there (*vide* Evidence, Rivers Conservancy Commission, 1877), and the quantity was 24,107 cubic feet per second, or 1,446,420 cubic feet per minute. He does not state the watershed area, but measuring this in like manner to that above mentioned, there appears to be as nearly as can be thus made out 2,000 square miles, or 1,280,000 acres, which gives 1·13 cubic foot per minute per acre. The distance down the valley is about 90 miles.

On the Nene at Peterborough, where the drainage area is, according to Mr. Beardmore's tables, 620 square miles, or 396,800 acres, Mr. W. Shelford stated to the British Association at the Dublin meeting in 1878, that a great flood there amounted to 8,000 cubic feet per second, or 480,000 cubic feet per minute, which would be 1·2 cubic foot per minute per acre of the drainage area. This is at about 50 miles below the extreme watershed.

On the Witham, Sir John Hawkshaw reported to the Drainage Committee in 1877, that the area draining into the river at Boston (at the Grand Sluice) is 504,000 acres, and that the quantity of water in floods to be provided for during periods of continuous rainfall would be 318,000 cubic feet per minute, or ·631 cubic foot per minute per acre. This is at about 60 miles below the extreme watershed.

On the Shannon, at Killaloe, it appears, from a paper read by Mr. James Lynam before the British Association in 1878, that the area of the catchment basin is 4,000 square miles, or 2,560,000 acres, and Mr. Robert Manning said that the maximum discharge there was 1,600,000 cubic feet per minute. This would be ·625 cubic feet per minute per acre. The distance from the extreme watershed is 140 miles.

Mr. Bateman said of the same river, before the Water Supply Commission, that he had had occasion, in investigating the inundations there, to ascertain what was the maximum flood to be provided for in all the rivers in the basin of the Shannon, and from information which he received from Mr. Forsyth, the engineer of the Board of Works, who had paid much attention to the subject, and from his (Mr. Bateman's) own measurements of the maximum floods in several parts of the Shannon, he found that 1 cubic foot per minute from each acre is a very considerable flood, and that there is no river falling into the Shannon, even when the volume of the flood is not checked by any lakes, which very materially modifies the amount of flood. There is no river which sends down more than 1·54 cubic foot per minute from each acre of ground.

Mr. James Dillon read a paper at the meeting of the British Association before mentioned, on works he had carried out on the Upper Inny river, where the area of the catchment basin is 273 square miles, or 175,000 acres, in which he said that he had measured floods there of ·4896 cubic feet per minute per acre, being nearly the same as the flood gauged by Mr. Unwin, at Windsor.

In the Report of the Commissioners appointed to inquire respecting the drainage of the district traversed by the river Barrow and its tributaries, in Ireland, in 1885, Mr. Robert Manning gave evidence of the maximum quantity of water flowing from very large areas during floods, as follows: Lough Neagh, 411,520 acres, 0·51 cubic foot per minute per acre; Woodford River, 101,455 acres, 1·12 cubic feet; Brosna River, a tributary of the Shannon, 285,000 acres, 0·72 cubic foot per minute per acre in the year 1876, 0·83 cubic foot in 1852, 0·93 cubic foot in 1851, and from that time to 1885 there had been no greater flood than this.

In the Ballinamore and Ballyconnell district, 90,000 acres, the quantity was 1·12 cubic feet per minute per acre, and from a small area of 3,200 acres in the Dunmoran district, county Sligo, the quantity was 3·60 cubic feet.

The largest estimate I remember to have been made of the probable quantity of water which might be produced by a maximum flood at or near the outfall of a great river is that made by Mr. Bateman in respect of the Severn at Gloucester, which was 2 cubic feet per minute per acre of the drainage ground.

There is a reservoir at the head of the river Churnet, in North Staffordshire, which was constructed solely for the purpose of compensation to mills for the abstraction of springs elsewhere which formerly flowed into the Churnet.

The watershed area is 6,500 acres, and notwithstanding that all the water passes through the reservoir, the waste weir was made only 60ft. long, while the greatest depth of water was but 40ft., and the area of the reservoir about 40 acres.

It was, therefore, soon filled up, and while it was full, on the 7th of August, 1862, a flood occurred, which rose upon the waste-weir to the height of from 4ft. 6in. to 5ft.

This, of course, is a depth far beyond any depths which have been experimented upon for the purpose of ascertaining the quantity from the observed depth; but as the weir had a clear drop of from 1ft. to 2ft., and the waste-water course sloped away from the weir rather steeply through the end of the bank, it may not be inappropriate to the occasion to apply a co-efficient of ·5 in the equation.

$$Q = c \times 5\tfrac{1}{3} \sqrt{d} \times l d$$

in which

$c = $ ·5, the co-efficient,
$d = 4\cdot5$, the depth of water,
$l = 60$, the length of the weir,
and $Q = $ the quantity in cubic feet per second.

This being so,

$$Q = \cdot 5 \times 5\tfrac{1}{3} \times 2\cdot12 \times 60 \times 4\cdot5 = 1{,}526.$$

To this must be added the quantity going through two outlet pipes, one 3ft. diameter, and the other 18in., which together would discharge 250 cubic feet per second, making in all 1,776, or 106,560 cubic feet per minute.

The drainage area is more than 6,500 acres; but as there were within the area, above the reservoir, a small reservoir and two mill-dams, which together had belonging to them about 500 acres, the area to be taken for the flood-ratio should be 6,000 acres, and dividing the number of cubic feet per minute, as above stated, by this area, the quantity would be 17·76, or, say, 18 cubic feet per minute per acre.

Thus a great flood, reckoned per acre of the ground from which it flows, varies much between the upper and lower portions of a river-basin, if in the latter case the whole river be gauged; but for any separate smaller area near the outfall the quantity per acre would, of course, be much greater, and might to some extent resemble that proceeding from an equal area in the upper part of the basin, in some rivers, but not in all, as may be seen in the two diagrams.

In those characteristics which influence the magnitude of floods, no two rivers of nearly the same length could differ more than the Thames and the Severn, in the upper-third portions of the watershed areas above the tideway, although for the lower two-thirds the respective falls are not very dissimilar, being in the Thames rather less and in the Severn rather more than 2ft. per mile for 100 miles above the tideway, as may be seen from the two diagrams

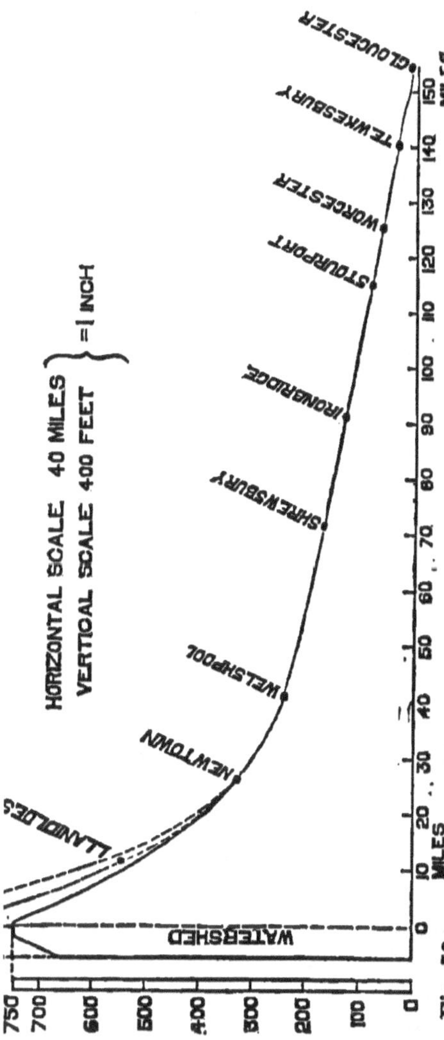

Fig. 56. LONGITUDINAL SECTION OF THE SEVERN BASIN

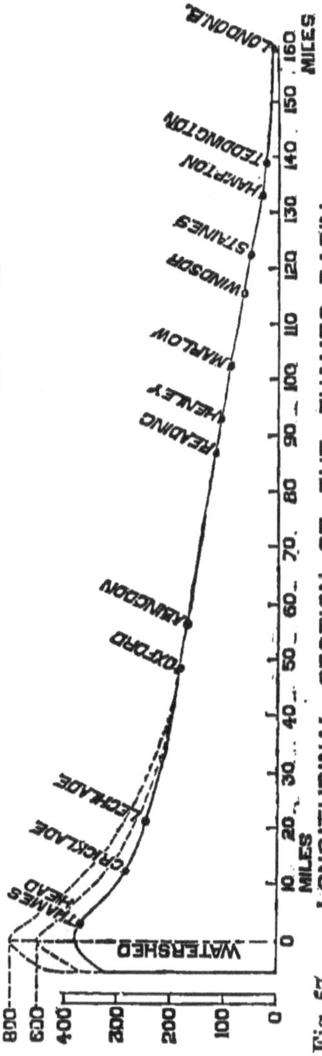

Fig. 57. LONGITUDINAL SECTION OF THE THAMES BASIN

here given (Figs. 56 and 57), the scales being the same in both.

Referring to diagram No. 56, if a position be taken up at a point either on the upper branch of the Severn, or on the Vyrnwy, within an average distance of 10 miles from the watershed, it will be at a height of about 500 ft. above the sea-level.

The watershed is, on the one hand, 1,550ft. above the assumed point of observation; and on the other, 1,250ft.; and between these two great general heights of watershed there is a pass which is 250ft. above the assumed point; and the length round that watershed is more than 50 miles, and its average height may be about 1,100ft. above the point mentioned.

Thus, then, the surface is very steep, and the rain-water descends in streams which fall 500ft. or 600ft. in the first 10 miles; in the second 10 miles, 200ft.; in the third, 120ft.; in the fourth, 60ft.; in the fifth, 30ft.; and forwards to Worcester, a distance of 80 miles or so, the fall is a little more than 2ft. per mile: that is, the general rate of inclination of the valley, but the river itself varies in different parts of its course both above and below that average, having in some parts accumulations of gravel and other *débris*, and in others hard benches of rock across its bed.

The Thames falls but 110ft. in the first 10 miles, taking a general average of its branches; 67ft. in the second 10 miles; 37ft. in the third; 22ft. in the fourth; and forward to Teddington the average fall is somewhat less than 2ft. per mile.

The ground in the upper part of the Severn consists of the hard impervious Silurian rocks.

The Thames comes from the Oolite and other pervious grounds in the upper part of the basin, running over a tract of Oxford clay, and then crossing the outcrops of the greensand and chalk, the area of these permeable strata being two-thirds of the whole area of the basin above the tideway; and as they absorb a great deal of the rain-water and give it out gradually during dry seasons, the stream is comparatively strong all through the summer.

The Severn is soon up and soon down, but in the Thames there is a more even flow.

It is very important to regard floods with reference to their frequency, as well as their magnitude. A few examples of maximum floods are given above, but they are such as occur very seldom, and mostly in winter, while floods of less magnitude and greater frequency do more harm in the long-run.

Mr. James Lynam said (British Association, Section G, 1878) of the Shannon floods, that the small floods had occurred every year and in every month. They kept the land saturated and cold during March, April, and May, which prevented the growth of good grass, and promoted the growth of sedge and weeds. The herbage grows only late in the season, and is late coming to maturity, and the mowing of the crop is thrown back into the rainy season.

In the large flat meadows there are three qualities of land, one letting at £2 to £3 an acre Irish, which would be at 24s. to 36s. an acre English, the second letting at £4 to £5 per Irish acre, the third letting at £6 to £8 an acre. (The Irish acre = 1a. 2r. 19p. English.)

The difference in the values of the meadows results, not from any difference in the soil itself, but from a difference of the levels of the lands.

The lands of the highest rents are 9in. to 15in. higher than the others, and are above water a month earlier in the spring. The sedgy, weedy meadows are the lowest by some inches, and are saturated longer than the others.

The kindness of the soil of these is evinced in many places, for on examining it closely numerous plants of clover, and some fine species of grass are seen, healthy, but small and distant; and if these lands were freed from saturation during spring, summer, and autumn, the clover and fine grass would flourish and extend.

Mr. Brundell,[*] in his evidence before the Rivers Conservancy Commission in 1877, said that on the river

[*] B. S. Brundell, Esq., M.Inst.C.E.

Don the works he had carried out there would not prevent the land being submerged by such a flood as occurred in July 1872, but it was safe against the more injurious summer floods.

There is no necessary connection between the storage of water in reservoirs and the prevention of floods in the sense that on the accomplishment of the one the other must necessarily follow, because reservoirs cannot be made large enough to effect that wholly; it must, indeed, be effected chiefly by the lowering of the water-level of the river; but nevertheless there is a dependence of one on the other to a certain extent.

Neither is there any practical reason why a river may not carry off flood-waters and yet be navigable, in cases where both purposes are desired. Duality of purpose is not new in engineering works, but rather it is of every-day occurrence.

SECTION XXVII.

STORAGE OF FLOOD-WATERS.

To stand in the middle of an extensive valley, by the river, and consider how the injurious effects of floods can be prevented, any reservoir for that purpose would seem to need to be preposterously large, to have any effect there. But that is not the point from which the question should be regarded; it would be beginning at the wrong end; rather, let every thousand acres downwards from the watershed be considered separately, or at most in areas of a few thousand acres together, and making sure provision for the floods of each, proceed step by step towards the middle and lower portion of the valley.

It is said that in France this question has been discussed since the year 1856 with regard to the valleys of the Seine, the Rhone, the Loire, and the Garonne (*vide* Foreign Abstracts, Minutes Proc. Inst.C.E., vol. lxvi., "On the Insufficiency of Reservoirs for Diminishing the Danger of Floods," by M. Gros), and that in the case of the Garonne a reservoir capacity of 720,000,000 cubic yards would be required "to protect Toulouse" from a flood similar to one which occurred there in 1875, and the conclusion is that the reservoirs which had been proposed for providing against floods in these vast areas must be abandoned. No doubt the French engineers have sufficient reason for coming to this conclusion, on this scale of things.

The Seine, as appears upon a geographical map of France, is 400 miles long, and drains 30,000 square miles; the Loire is 530 miles long, and drains 45,000 square miles; and the Garonne is 300 miles, draining 32,000 square miles, and the Rhone 35,000 square miles in France besides a further tract in Switzerland; indeed, these four rivers appear to drain about two-thirds of the whole country.

But for the purposes of river-conservancy in England the requirements are less grand. Here we have in England and Wales but 58,186 square miles altogether, divided into 211 distinct watershed areas or river-basins. Many of these, however, are not injuriously affected by floods. The number is made up by 54 draining 7,046 square miles towards the south coast, being an average of 130 square miles; 30 draining towards the east coast, an area of 16,084 square miles, the average being 536; 29 draining towards the north-east coast, an area of 13,291 square miles, being an average of 457; 17 towards the north-west coast, draining an area of 2,345 square miles, or an average of 138; and 81 on the west coast, draining 19,029 square miles, or an average of 236 square miles each.

Even these are large areas, and a thousand or two thousand acres is a very small part of some of them, and the regulation of the flow off the ground of excessive rainfalls on any such area would do but little to mitigate the effects of floods in the lower part of such large areas. But it stands good for its own position, and if it deals with floods upon its own area it is an effectual work, so far; and the effect may be extended downwards to any distance to which favourable sites extend.

Where the provision for regulating floods can be combined with a useful reservoir the expense would be very much less.

An example has been given of a reservoir and of a regulating dam, each made separately, the reservoir to contain 46,000,000 cubic feet for a drainage area of 2,000 acres where the average annual rainfall is 33in., and the dam to contain 12,000,000 cubic feet per thousand acres of the drainage area irrespective of annual rainfall.

Let an example now be taken in which the two are combined: The rainstorm producing 4in. in depth of water, of which a quantity equivalent to 3in. in depth

CUBIC CONTENTS OF EMBANKMENTS.

might flow into the regulating dam in a short time, was taken in the preceding examples as proceeding from 1,000 acres of ground; but it might extend over the whole 2,000 acres taken as the drainage area in the example of the reservoir, and in that case 24,000,000 cubic feet of water would have to be provided for in addition to the 46,000,000 of the reservoir, making together 70,000,000 cubic feet. This would be impounded by an embankment 70ft. high.

The quantities of earthwork in embankments of different heights on the same site are proportionate to the cubes of the heights, if to the height of the bank itself be added a height equal to the width of the top of the bank divided by the ratios of the two slopes added together; thus, if the slopes are 3 to 1 and 2 to 1 respectively inside and outside, and the top width of the bank is 20ft., $\frac{20}{3+2} = 4$ft. to be added to the actual height of the embankment to find the apex in which the prolongation of the two slopes would meet, and to which point the

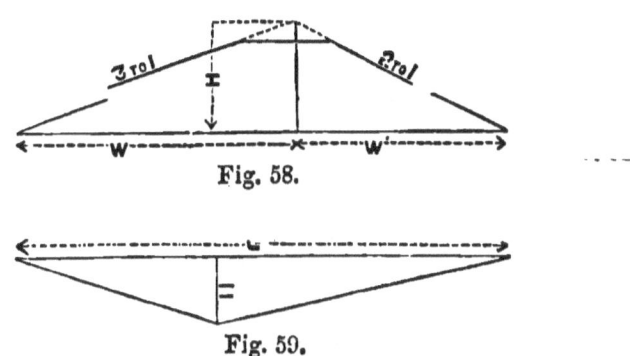

Fig. 58.

Fig. 59.

height is to be measured. Fig. 58 being a cross-section, and Fig. 59 a longitudinal section of an embankment, the cubic contents are—

$$\frac{L \times H \times \overline{w + w^1}}{2 \times 3}$$

In many embankments some part of the ground towards

the middle of the bank is more or less flat, but for the purpose of comparing the quantities of earthwork in embankments of different heights, the side slopes may be assumed to meet in a point, somewhat lower than the actual surface of the ground at the middle of the bank. If there are on a particular site 80,860 cubic yards in an embankment 60ft. high above this point, there would be in one 70ft. high—adding in each case 4ft. for the apex— $(64)^3 : (74)^3 :: 80,860 : 125,000$ cubic yards, and, deducting the puddle wall, the embankment would be 116,000 cubic yards, for a storage-room of 46,000,000 cubic feet, with the addition of a space of 24,000,000 cubic feet to receive a flood from 2,000 acres.

The water coming into the 46,000,000 cubic feet space would be discharged by the same valves and in the same manner as has been described, the only difference being that the valve-tower would be 10ft. higher, and the water which would rise into the upper space of 24,000,000 cubic feet would be discharged by a separate sluice, nearer the end of the bank, at the level at which the waste-weir would otherwise be laid, and by the same means as was before described for a separate dam, there being for this discharge a separate culvert; and it might advantageously be placed near that end of the embankment at which the bye-wash would be made, and might discharge into it.

The sill of the sluice would be laid at a depth of 10ft. below the assumed top-water level of the flood space, leaving the lower space, or reservoir proper, intact; but the invert of the culvert should be laid a foot lower. In the other case the height of the sluice-opening was described to be greater than the width; in this one the width should be the greater, and it may conveniently be made twice the height.

In this reservoir there would be a depth of 10ft. above the level, which would otherwise be the top-water level, to contain a flood, instead of its running away over the waste-weir otherwise provided at the height which allows 46,000,000 cubic feet of water to accumulate before any passes over it. The top-water level is 10ft. higher, and the top of the bank is 6ft. higher still.

This provision of 24,000,000 cubic feet of space for a flood stands instead of the provision of a waste-weir and bye-wash, which, logically, are unnecessary. But to run no risk let them be provided, although they will not be required, in all probability. But they need not be so capacious as if no such flood-space were provided. Let them be of half the capacity. The estimate would then stand thus:—

	Quantity.	Price. s. d.	£
Clearing the seat of the embankment..	15,600 sq. yd.	0 2	130
Excavation of the puddle trench.. ..	13,000 c. yd.	5 0	3,250
Clay puddle	22,000 ,,	2 6	2,750
Embankment	116,000 ,,	1 6	8,700
Road on top of embankment	1,000 sq. yd.	2 0	100
Covering the inner slope with stone ..	9,000 ,,	2 0	900
Soiling the outer slope	7,200 ,,	0 2	60
Excavation for culverts and valve-tower	5,000 c. yd.	4 0	1,000
Ditto for bye-wash	2,000 ,,	1 0	100
Concrete in foundations..	500 ,,	6 0	150
Ditto in bye-wash	1,200 ,,	7 0	420
Brickwork in culverts and valve-tower	900 ,,	24 0	1,080
Gravel puddle..	800 ,,	4 0	160
Stone sills, caps, coping, &c.	1,200 c. ft.	2 0	120
Valves, sluice, and other ironwork ..	.. ..	.. ..	300
Footbridge	.. ..	.. ..	200
Gauge-weir	.. ..	.. ..	100
			19,520
Contingencies			1,980
			21,500
Land and fencing			6,500
			£28,000

Adding to this a sum of £3,000 for expenses other than those of works and land, the cost of the combined reservoir and regulating dam in this situation would be £31,000, and deducting the cost of the reservoir alone, as found before—viz., £22,000—the cost of the provision for floods thus combined with a storage reservoir would be £4 10s. per acre, or equivalent to a yearly contribution, at 5 per cent., of 4s. 6d. per acre.

Thirty-three inches annual rainfall, adopted in the example of a storage reservoir, is not far from the average of the whole of England and Wales, but the average of the upper and middle portions of the river-basins is greater.

The work about which these general remarks are made can be practically dealt with in no other way than by taking each case as it arises with its attendant circumstances; but we may give one other example, with a more extended area and a greater rainfall, where the area is 3,000 acres or a little more, and the average annual rainfall is a third greater than that before assumed, *i.e.*, where it is 44in. Here the mean of three consecutive years of least rainfall would be about 36in., and deducting 12in. for evaporation, &c., the available annual depth is 24in. This yields, by means of a reservoir, a daily quantity the year round of 720,000 cubic feet. If one-third be discharged into the stream, there are left for supply 480,000 cubic feet per day, or 3,000,000 gallons.

Where there is this amount of rainfall there will probably be a greater number of wet days than where the amount is but 33in., and the number of days' supply the reservoir should hold would therefore be less. Let it be assumed that 150 days would be in this case the proper length of time during which the storage must be depended upon. This would indicate a reservoir capacity of 108,000,000 cubic feet.

If the same proportions be followed as before, this quantity would be contained in a reservoir having a depth of 72ft. at the site of the embankment; but let the depth be taken at 74ft., and the height of the bank at 80ft. The top width of the other banks is assumed to be 20ft.; in this one let the top width be 24ft., or if the 20ft. be retained, let the quantity of earth equivalent to the extra 4ft. be added in benches on the outer slope. The embankment would then contain, after deducting the puddle-wall, 174,000 cubic yards, and the estimate for this reservoir would be as follows :—

COST OF A LARGE RESERVOIR.

	Quantity.	Price. s. d.	£
Clearing the seat of the embankment..	21,600 sq. yd.	0 2	180
Excavation of the puddle trench.. ..	20,400 c. yd.	5 0	5,100
Clay puddle	32,400 „	2 6	4,050
Embankment	174,000 „	1 6	13,050
Road on top of embankment..	1,200 sq. yd.	2 0	120
Covering the inner slope with stone ..	13,000 „	2 0	1,300
Soiling the outer slope	9,600 „	0 2	80
Excavation for culvert and valve-tower	5,500 c. yd.	4 0	1,100
Ditto for bye-wash	6,000 „	1 0	300
Concrete in foundations..	550 „	6 0	165
Ditto in bye-wash	3,200 „	7 0	1,120
Brickwork in culvert and valve-tower..	1,000 „	24 0	1,200
Gravel puddle..	900 „	4 0	180
Stone sills, caps, coping, &c...	2,000 c. ft.	2 0	200
Valves, &c.	..	.. ..	270
Footbridge	..	.. ..	230
Gauge-weir	..	.. ..	200
			28,840
Contingencies			2,860
			31,700
Land and fencing			9,300
			£41,000

To this, for other expenses than those of works and land, let £4,000 be added, making £45,000. This, for a storage of 108,000,000 cubic feet, would be about £420 per million.

To complete the form in which we desire to put this question of water-storage combined with flood-regulation, let now 24,000,000 cubic feet of space be added to this reservoir, which is assumed to be of suitable capacity for an area of 3,000 acres, and an average annual rainfall of 44in., while at the same time it is assumed that one rain-storm in any district of a few square miles amounts to 4in. in depth over an area of 2,000 acres only; and that, of this depth, three-fourths run directly off the ground, making 24,000,000 to be added to 108,000,000, or together requiring a space of 132,000,000 cubic feet.

This would be impounded by an embankment 86ft. high, which would contain, after deducting the puddle-wall, 214,000 cubic yards, and the whole cost of this reservoir, with the addition of the flood-space above mentioned, would be £45,000 for works and land, to which may be added £5,000, making £50,000. This reservoir would equalise and render useful the streams of 3,000 acres of ground, and the addition of the flood-space would protect that area from any demand for contribution towards the funds of a river conservancy.

If the supply of water from the reservoir, viz., 3,000,000 gallons per day, pays for its construction and attendant expenses, the cost of this flood protection would be very small, being £5,000 for 3,000 acres.

SECTION XXVIII.

Relief of Land from Floods.

The question of relieving lands from floods ought, in the first instance, to be considered acre by acre downward from the top of every hill, for the purpose of seeing whether and to what extent the excessive rainfalls may be prevented from descending into the valley too suddenly, and in what manner that should be done, should it be possible to effect the object by any such means; and if not, to see what provision is necessary in the valley itself, by way of removing obstructions in a river-course, or enlarging it, or limiting the lateral spread of the waters by embankments, or by a combination of the two latter means.

The numerous deputations of landowners to the Local Government Board have set forth, and the same thing has been stated in Parliament, that it is a hardship to tax the upper lands in a river-basin for the benefit of the lower lands, or that portion of them liable to floods. The extent of land liable to floods is small in comparison with the whole area of a river-basin, and the objectors to these propositions have said that if improvements in valleys will pay, the owners of those lands will carry out the works required, and that if they will not pay, it is unjust to tax the larger area for the benefit of the smaller. That would be so if the proprietors of the upper lands were innocent of contributing to the floods of the valleys; but, from a consideration of the following circumstances, it would appear that they are not so. The bed of the river belongs to the proprietors of the adjoining lands, as far as the centre from each side.

It is curious, it may be incidentally remarked, that the bed of a river, which no one can see or touch, should belong to the proprietor of the adjoining land, but it may be so on the principle that it must belong to somebody, and, except the lord of the manor, there is no other person to whom it could belong, excepting also those cases in which it may have been transferred by Act of Parliament to another proprietary, and those in which, by another right, it belongs to the Admiralty Board, or to the Board of Trade, by transference to that department, that is, within the high-water mark of the tideway; but with these exceptions the river-bed belongs to the owners of the adjoining lands, and yet the proprietor of non-riparian lands uses the river-bed for the conveyance of water from his land to the sea, and that, too, without so regulating its flow as to prevent the lower lands being flooded by it.

It is proposed to divide the lands of a river-basin into three parts, lowlands, midlands, and uplands. Taking the uplands and midlands as one, they might amongst them contribute, say, one-half of the amount, the other half being paid by the flooded lands.

If in any river-basin one-tenth of the whole area were subject to floods, and if the cost of relieving that portion from floods were £9 an acre, the statement would be:—

	£	s.	d.
1 acre of flooded land	4	10	0
9 other acres	4	10	0
	9	0	0

The average amount per acre to be paid by the other lands would thus be 10s., which would be one-ninth of the amount per acre paid by the flooded lands, and at 5 per cent. interest it would be 6d. per acre per annum. If only one-twentieth of the whole area of a river-basin be subject to floods, the statement would be:—

	£	s.	d.
1 acre of flooded land	4	10	0
19 other acres	4	10	0
	9	0	0

And the average amount to be paid by the other lands would be a little less than 5s. per acre, or one-nineteenth part of the amount per acre paid by the flooded lands; and at 5 per cent. interest it would be 3d. per acre per annum. But it may be said that while this or that might be a fair average, acres differ from other acres in value, and that moorlands ought not to contribute so much as cultivated lands. If so, and no doubt it may be so, the nearest approach to an absolutely accurate basis is the rateable value of each description of land, the average of the whole amounting to the average rate required.

Neither would the flooded land be equally rated per acre; but in this case the proper basis would be the relative frequency of the floods to which the different parts are subject, as (1) to minor floods, which may occur every year; (2) great floods, which may not occur oftener than once in three years; and (3) the highest floods, which may not happen more than once in six years or a longer time. Where the records of the relative frequency of these floods in times past are not sufficiently numerous or accurate, an approximation by levels may be made, such as 8ft., 12ft., and 14ft., respectively, above the summer level of the river as a datum; and each level might contribute in something like the following proportions:—

	£	s.	d.	
The lowest level	6	0	0	per acre.
The middle level	4	10	0	,, ,,
The highest level	3	0	0	,, ,,
Average	4	10	0	,, ,,

But this could only be properly adjusted after relative areas of the three levels of any case had been ascertained; and indeed the whole statement is but an indication of data. Much labour is necessary to be undertaken in each particular case, before the true proportions can be found.

It is not, however, agricultural lands only which suffer from floods; towns suffer greatly. The proper basis upon

T

which they should contribute to the river conservancy fund would be probably the same as that for land, if they contribute at all; but the more likely thing for a borough, or a local board district, or the district of a town's improvement commission, is that the Corporation or the Board would themselves do what is necessary within their own boundaries to provide such a passage for the river as would injure no land above the town, that is, to make and maintain the water-level standard at the upper boundary of their district efficient for the drainage of all lands above that point.

It would hardly be practicable to rate a town for any purpose requiring the fund to be expended beyond the boundary; it would not indeed be possible without an alteration of the law in that respect, but even if this were done there would remain the difficulty of an amicable agreement between the town authorities and the general conservancy board on the proper contribution of the town.

As the free flow of a river through a town is much obstructed, it is undeniable that towns should contribute in greater proportion, if at all, and should at least keep down the water-level at the upper boundary so as not to flood the land above the town; but besides this, the water flowing from the town area increases the floods in the river at the lower boundary to a greater than an average degree per acre of the drainage area.

The water-supply brought into the town by other means and from another source runs into the river through the sewers; but its proportionate volume to that arising from a heavy rainfall is very small, being that of the dry-weather flow of the sewers only; but the same depth of rainfall upon an area built upon contributes a greater quantity of water per minute to the river than does an equal area of agricultural land, as it runs off more quickly, and therefore the town is justly liable to a proportionately greater rate of contribution per acre of its area.

The proper proportion would be, that between the rate

of flow off the ground from a given depth of rainfall on a steep hill-side of hard and impermeable rock, and that from the average of all the land in a river-basin, because, whether the town be hilly or flat, its surfaces throw off the water quickly; and if that proportion be regarded, it would be, say, as 18 (12 to 24 cubic feet per minute per acre) to $1\frac{1}{2}$ (1 to 2 cubic feet per minute per acre), or as 12 to 1.

But the municipal authorities would prefer to do at their own expense what is necessary within their own boundaries. Where, however, land is built upon in rural districts—as in the case of small towns in rural sanitary districts—the same principle is applicable, although the proportion may not justly be the same, because the ground is not so completely built upon, on the average of the town area, as it is in large towns, and therefore the volume of water contributed to a river from a given area under a given rainfall is proportionately less, and might, perhaps, be not unjustly stated at one-half that of the other case, and if so, the rate per acre of contribution to the river conservancy fund would be 6 to 1 of the agricultural land in the same river-basin above the flood-level. But perhaps the basis of rating towns which has been proposed by Mr. Bailey Denton[*] would more accurately meet the case.

In respect of the higher lands of a river-basin, however, the great question of all is whether some means of regulating the flow of the excessive rainfalls off the ground cannot be adopted, which would change the whole question from one of aid to the lower lands to one of absolute benefit to the proprietors of the upper lands; whether, as the rain first descends upon them, they may not be enabled to exercise their prior right in the possession of the water; and if the adoption of such means should be of no greater benefit than would be equal to the liability to contribute to the freeing of the lower lands from floods,

---

[*] J. Bailey Denton, Esq., M.Inst.C.E.

it would at least dispose of the difficulty of terms of contribution. But, in fact, it would do a great deal more.

A large river-basin contains within it hills, not usually so high as the main watershed, but still of considerable height, along the tops of which lie the watersheds of the minor rivers and streams. The ground is as various in character as the several geological formations—some hard and igneous; some which, although of sedimentary origin, have been changed in character by contact with igneous rocks; some more slaty and others containing more clay; others again of limestone, and some of sandstone and shale; some of chalk, oolite, and lias.

Accordingly the hill-sides of a river basin are various in the character of the surface; in one case it is pretty even and descends to the valley almost in one unbroken slope; in another it is broken up into ravines; and in others the chief feature of the river basin is a number of tributary streams on which proper sites for storage reservoirs exist. It has been strongly insisted upon that it is impossible to prevent floods in rivers by the construction of any reservoirs whatever, but that cannot be wholly assented to, although for the larger part of the area of any river-basin it is true.

For instance, in 1,000 square miles—to go no further, although some of the river-basins are much larger than that—there are 640,000 acres, and assuming 10 per cent. to be subject to floods, there would be 576,000 acres above flood-level; but it is only the higher parts of this area which could be dealt with by storing or regulating the water flowing from them; that from the larger portion of the area must continue to flow directly to the river and larger streams. But, looking broadly at the question, the river may be said to be equal to its duty of carrying off the water which comes to it, and it is only the top of the flood, as one might say, which needs keeping back for a time until the river has begun to subside, and it may not be impossible to do this on the higher portions of the area, though not necessarily, for this purpose, on the highest, as is the case in waterworks reservoirs.

When an engineer looks for a site for a storage reservoir, or perhaps for several reservoirs, for the supply of a town, the elevation must be sufficient, and in general must be great, the best water being usually found at the greatest distance, and a great distance requires a correspondingly great fall in the conduit; and as for economy, there must be only one conduit to supply all parts of the town, high as well as low; the choice of site for a reservoir is therefore very limited, and is usually found only within some few thousand acres of the upper part of a river-basin, and not far from the main watershed.

But the same difficulty does not arise where the object is merely to control the flow of a bulk of water so that its discharge may be gradual, and prevent the river overflowing, for as relative elevation need not be regarded, a wide range is afforded for the choice of sites for reservoirs; and, moreover, control of the water may be managed in a different way from that of the construction of reservoirs properly so called, in those river-basins where the geological formation and the character of the surface are favourable.

There are two respects in which the regulation of the flow of excessive rainfall off the ground is to be regarded: that of absolute storage for a considerable length of time, and that of merely regulating the flow, or controlling the water until the flood in the river has begun to subside; and we will offer a few remarks upon the practicability of this, presuming that the owner of these high lands will have—if, indeed, already he has not—power to make such charges for the use of water as will be sufficiently remunerative, or at least will be equivalent to his liability in the matter contemplated by one or two Bills which have been before Parliament.

SECTION XXIX.

Regulation of Flood Waters.

In the year 1877 an Act was passed for England and Wales, and for Ireland, which gives facilities to landowners in the construction of reservoirs and similar works, on the ground that " in many places it would greatly conduce to the affording of a plentiful supply of water to the inhabitants of villages and towns, and to the industrial requirements of the locality, if facilities were given to landowners of limited interests to charge their estates with sums expended by them in constructing reservoirs and other works for the supply of water, of a character permanently to increase the value of such estates for other than agricultural purposes;" and with this Act is incorporated the Waterworks Clauses Act, 1863, with respect to the security of reservoirs, and also the Improvement of Land Act, 1864; and the supply of water from reservoirs so constructed to sanitary and other local authorities, and to manufacturers and other persons, is deemed in this Act an improvement of the land within the meaning of the Improvement of Land Act of 1864, if the construction of such works will effect a permanent increase in the value of the land, or produce a greater income; and there would be this further advantage to land within the watershed area appropriated to any such reservoir, that in respect of any contribution to funds for the prevention or mitigation of floods lower down the valley, the watershed area so appropriated would properly be exempt, the reservoir being so constructed as to either store or regulate the flow of flood waters from its watershed area.

In connection with this subject, numerous instances of

heavy rainfalls may be seen in Mr. Symons's records of rainfall as follow:—In one day, 3in., 3·12in., 3·31in., 3·34in., 3·42in., 3·47in., 3·55in., 3·65in., 3·70in., 3·75in., 3·80in., 3·90in., 3·92in., 4·20in., 4·24in., 4·27in.; so that we may say 4in. depth of rainfall may occur in 24 hours, being at the average rate of $\frac{1}{6}$in. per hour.

But a greater depth per hour often takes place during shorter periods of time; as, for instance, ·27in. in $3\frac{1}{2}$ minutes, ·31in. in 5 minutes, 1·48in. in 20 minutes, ·64in. in 30 minutes, ·75in. in 1 hour, 1·53in. in $2\frac{1}{2}$ hours, 2·60in. in 3 hours, 3·60in. in $4\frac{1}{2}$ hours; and again, ·89in. in 10 minutes, ·68in. in 15 minutes, 1·25in. in 30 minutes, 1in. in 40 minutes, 1·39in. in 45 minutes, 1·78in. in 1 hour, 2·10in. in $1\frac{1}{2}$ hour, 2·61in. in 2 hours, and so on; and it was remarked in 1878 by Mr. Symons, that in every year there is sure to be a fall, somewhere in England and Wales, of at least ·30in. in 15 minutes, or $\frac{1}{2}$ in. in 30 minutes, or ·60in. in 45 minutes, or ·70in. in 1 hour, or ·80in. in 2 hours.

Referring to what has already been said of the quantity of water running off the ground, as proved by stream-gaugings, assisted sometimes by approximate estimates of the quantity derived from other measurements, 30 cubic feet per minute per acre of the watershed area has been found to run off, or 500 cubic feet per second from 1,000 acres, if the rainstorm extends over that area. This is equivalent to $\frac{1}{2}$in. depth of water over the ground running off in an hour, and exceeds the ·80in. in 2 hours mentioned by Mr. Symons; and it exceeds it, moreover, in another respect, inasmuch as though this 30 cubic feet per minute per acre is the quantity running off the ground, it would probably fall in less time. Rainwater never does, in fact, run off as fast as it falls; nor does it extend over any large tract of ground with the intensity indicated in the examples mentioned.

Where the ground is of a porous nature, as some of the stratified formations are, it may become saturated with water during long-continued wet weather before the heavy rainstorm begins, and yet even in this case it does not run off at the same rate as that with which it falls; for where the ground does become saturated in this way it is comparatively flat, and the depth of water may be seen during the rainstorm to increase where it falls, which is a fixed proof of the proposition; and even where the ground consists of hard, bare, and precipitous rocks, and the surface consequently is steep, the time of falling of the rain is not so long as the time of running off.

In taking therefore $\frac{1}{2}$in. in depth as the quantity running off in an hour, at the rate of 30 cubic feet per minute, it may be regarded as representing a rainfall of more than $\frac{1}{2}$in. in an hour. The comparative times on different kinds of ground and forms of surface cannot be stated from exact observations for every kind of ground in a general form, but an approximation is two-thirds of the rainfall running off in the time in which it falls, in excessive cases such as those mentioned, so that where $\frac{1}{2}$in. in an hour is found by gaugings to run off the ground it would probably represent a rainfall of $\frac{3}{4}$in. in that time. But as in floods it is extreme cases which must be provided for, let the case be calculated on the basis of 1in. of rainfall in an hour running off the ground at the rate of $\frac{3}{4}$in. per hour, or at the rate of 40 cubic feet per minute per acre of the area.

The question then arises, how much of this during the same time will the river carry off without overflowing, after the shoals which have accumulated have been removed, and in some cases benches of rock across the river's bed have been excavated, and other unnecessary obstructions removed?

The difference between these two quantities is that which would be dealt with by storage or regulation; that is by storage when there is or may be by storage created a demand for water for power, or in the other case by regulation only where there would be no such demand. If a river be examined and its capacity for carrying off flood waters be ascertained, the quantity so found deducted from that which proceeds immediately from the rainfall is the quantity to be dealt with by storage or regulation as the case may be. Such observations as I have been able to make point to the probability that, approximately, the river will carry off 20 cubic feet of water per minute per acre of the ground over which these heavy rainfalls extend, and where that is so 20 cubic feet per minute per acre would accumulate in the reservoir, or in the regulating dam, until the river-flow had begun to subside, or had subsided to a given level.

It is easy to fix this level by observation of the river in each case, and when fixed upon the outlet valves can be made to open and close, to any degree required, by means of the action of the rise and fall of the water itself, and the contents of the regulating dam can be discharged in any extended time desired, the discharge commencing when the river is in a proper condition to carry off the water.

The capacity of a reservoir or of a regulating dam proper for any given situation must be determined on the basis of a given quantity of water flowing from a known area of ground, as an acre or a thousand acres, constituting a part or the whole of the area appropriated, or from which the water to be dealt with flows; and that quantity varies with the position in which works may be constructed.

A reservoir for the purpose of storing or regulating the flow of water off the ground may be made either on the extreme upper part of a river-basin—that is, within the few thousand acres contiguous to the main watershed—or it may be made lower down towards the valley; wherever, indeed, the most favourable sites exist, although these will seldom, if ever, be far down the valley; for a favourable reservoir site is one in which the opposite hillsides approach each other at a spot where an embankment may be made, and where the ground above the site of the embankment widens out, so that with a comparatively short bank a large quantity of water may be impounded.

The variable quantity of flood water per acre of the watershed area, according to the position of the place of observation with respect to its being in the upper or the lower portion of a river-basin, shows that the cubic capacity of the reservoir per acre of its watershed area must be different in the two cases, notwithstanding that the same depth of rain may fall in either position for a short time; and the cause of difference between the two cases is that in the one the heavy rain may extend over the whole watershed area, when it is not very large, while in the other it would extend over part of the area only. This is found to be so by actual observation within the areas of 5,000 to 8,000 acres belonging to reservoirs in the upper parts of river-basins, for where there are several streams within such an area, divided by subordinate watersheds, it is frequently observed that one of the streams is much more swollen than another.

The sites of reservoirs, formed by making an earthen embankment across the stream from one hill-side to the opposite one, vary much in respect of the quantity of water capable of being impounded by embankments of the same height, and in the area covered with water.

The following are examples of the actual capacities of different sites. Taking three different reservoir-sites which vary amongst themselves about as much as many others do in respect of holding-capacity with various heights of bank, and ascertaining their capacities at each of the top-water levels of 20ft., 28ft., 36ft., 44ft., 52ft., and 60ft. above the seat of the embankment in each case, the following capacities with various heights of bank appear. (*See* next page.)

The track of a great rainstorm such as those producing 3in. depth of water in a day, or even 4in., may extend to 3 miles in length and a mile in width. This would cover nearly 2,000 acres of ground; and let the depth actually running off from it be 3in., which would be an extreme case. The quantity due to such a depth over that area would be $2,000 \times 43,560 \times \cdot 25 = 21,780,000$ cubic feet; but let the capacity of the regulating dam be 24,000,000 cubic feet.

EXAMPLES OF RESERVOIR-CAPACITY. 283

That would be at the rate of 12,000 cubic feet per acre of the ground on which the rainstorm falls; and where the regulating dam is not combined with a storage reservoir for other purposes that seems sufficient, for if it provides for the reception of a sudden rainfall of 3in. over 2,000 acres of ground, it may be reckoned as pretty certain that no second rainstorm will succeed one of that intensity

	Area.	Capacity.
	acres.	cubic feet.
Height of embankment, 24ft. . Greatest depth of water, 20ft.	16¼ 23¼ 31½	6,750,000 10,500,000 15,750,000
Height of embankment, 32ft. . Greatest depth of water, 28ft.	21½ 34½ 41½	13,250,000 21,000,000 28,500,000
Height of embankment, 41ft. . Greatest depth of water, 36ft.	26¼ 47¼ 55¼	21,500,000 34,500,000 45,500,000
Height of embankment, 49ft. . Greatest depth of water, 44ft.	34 59 76¼	32,000,000 53,000,000 68,500,000
Height of embankment, 57ft. . Greatest depth of water, 52ft.	44½ 75½ 94½	46,000,000 75,250,000 98,250,000
Height of embankment, 66ft. . Greatest depth of water, 60ft.	49¼ 96½ 110½	62,500,000 105,500,000 134,000,000

until long after that one has been discharged by the river.

Such a rainstorm may occur in any portion of the upper part of a river-basin, and therefore it may be said that reservoirs would require to be numerous to meet all contingencies, and that, no doubt, is so; they would require, indeed, to be made on every favourable site, whether it be favourable for a large or a small reservoir; but for

2,000 acres it would be as above stated. In situations, however, where a reservoir may be constructed for other useful purposes on the basis of storage of the average annual rainfall of three successive dry years, such as that of 46 million cubic feet for a watershed area of 2,000 acres where the average annual rainfall is 33in., and where 180 would be the proper number of days' storage, one reservoir may be made to serve both purposes, by providing in addition to the capacity of the ordinary reservoir a space of 24 million cubic feet. In this case, from the ordinary reservoir the water would be discharged near the bottom, or in any case below the top-water level, and from the upper space or regulating reservoir-room it would be discharged at the top-water level of the ordinary reservoir through an opening which would discharge a given maximum quantity per minute, whatever height the water may rise to, so that in regularity of discharge it would act automatically. As the surface of the water in the ordinary portion of the reservoir would vary in altitude below a certain height, a heavy rainstorm may amount to no greater quantity than will fill it up to its standard height, but when this portion is full, or nearly so, the water of a heavy rainstorm will rise into its own space and be discharged by its own outlet.

With respect to the economy of constructing reservoirs in such situations as those indicated, some of which might be for purposes of water-power or for water-supply to the villages near them and to towns farther off, while some might be solely for the purpose of regulating the flow of water off the ground and preventing floods so far as they might command considerable watershed areas, it should be said also that in the year 1875 a large area of land in the midland counties of England was flooded by a depth of from 1in. to 2in. of rain falling continuously during two or three days—amounting, that is to say, to that depth in that time. Such a state of things could not, of course, be much affected by the construction of any reservoirs of practicable dimensions, such as those mentioned, but it need not therefore be concluded that such reservoirs would be useless; they would be perfectly successful for the purposes named, and as far as the ground appropriated might extend in each case.

It comes to this, that where a reservoir is calculated on the basis mentioned, viz., to store and distribute day by day an equal portion of the rainfall of three successive years of least rainfall, an additional space equal to that of one heavy rainstorm within the watershed area of the reservoir should be provided to receive it, hold it temporarily, and discharge it into the river when the flood below has so far subsided as to enable the river to carry off the water.

A rainstorm 3 miles in length and a mile wide, covering nearly 2,000 acres of ground, and of such intensity that a depth of 3in. of water over the whole of that area proceeds from it, has been taken as an extreme instance to be provided for, and which would be provided for by a space of 24 million cubic feet. If in this situation a reservoir for the storage and equalisation of the rainfall of three successive years, as before mentioned, were made where the average annual rainfall is 33in., and where 180 days' storage should be provided, the capacity of the reservoir would be, say, 46 million cubic feet, and would be found at one of the sites mentioned with an embankment 41ft. in height, the greatest depth of water being 36ft., and covering an area of 56 acres; or it would be found at another of the sites with an embankment 46ft. in height, the greatest depth of water being 41ft., and covering an area of 56 acres; or at another of the sites with an embankment 57ft. in height, the greatest depth of water being 52ft., and covering an area of 45 acres.

If to this ordinary reservoir there be added 24 million cubic feet of space for a regulating dam, the whole capacity would be 70 million cubic feet. This would be found at one of the sites already mentioned with an embankment 50ft. in height, the greatest depth of water being 45ft., and the area 78 acres; or it would be found at another of the sites with an embankment 55ft. in height, the depth of water being 50ft., and the area 73 acres; or at another with an embankment 70ft. in height, a depth of water of 64ft., and an area of 53 acres.

In the case of a larger area, say 5,000 acres, and with the same annual rainfall, 33in., and a loss by evaporation of 14in., the daily average yield would be about 700,000 cubic feet, and where 150 would be the proper number of days' storage, the capacity of an ordinary reservoir would be 105 million cubic feet. This would be found at one of the sites referred to with an embankment 66ft. in height, the greatest depth of water being 60ft., and covering an area of 96 acres; and if in addition to this 24 million cubic feet of space be provided as before, the whole capacity would be 129 million cubic feet, and such a reservoir would be found at one of the sites mentioned with an embankment 65ft. in height, the greatest depth of water being 59ft., and the area covered 110 acres. These instances from actual works may indicate approximately what would probably be the conditions in other cases, but of course every case must be taken actually according to its own circumstances.

SECTION XXX.

RIVER CONSERVANCY.

Two papers on the conservancy of rivers are printed in the March volume of the 'Minutes of Proceedings' of the Institution of Civil Engineers, 1882, one by Mr. Wheeler,* of Boston, on the fen rivers of the east coast, and the other by Mr. Jacob,† of Salford, on the river Irwell. The circumstances of the rivers treated of in these two papers are widely different, but they do not between them embrace all conditions of rivers. Those dealt with by Mr. Wheeler are the Witham, Welland, Nene, and Ouse, draining the eastern midland portion of England lying between the Trent, the Severn, and the Thames. They are typical of the drainage systems of flat districts of permeable strata, with a small rainfall, and discharging into sandy estuaries, like the Wash. These rivers are still subject to most disastrous floods, although large sums of money have been expended upon them, because no river has been considered as a whole, but each dealt with in part only, and not from the outfall upwards.

There are other smaller rivers draining the district lying between the Ouse and the Thames, which Mr. Wheeler's paper does not deal with. The area drained by the four principal rivers is 5,719 square miles, their total length being 416 miles, and if the tributaries be included the total length is 872 miles. The principal towns within the watershed of these four rivers are

---

* W. H. Wheeler, Esq., M.Inst.C.E.
† Arthur Jacob, Esq., M.Inst.C.E.

Lincoln, Boston, Grantham, Spalding, Wisbech, Peterborough, Northampton, Lynn, Cambridge, Ely, Bedford, and Dunstable. The average rainfall is 26·05in., the least depth recorded being 17·39in., and the greatest 34·48in.

We have said that if rainfall records of long periods of time be examined, the maximum would probably be found to be about twice as much as the minimum fall, the one being about a third greater than the average, and the other a third less. How nearly that rule applies in this case may be seen in the Appendix No. 1 to Mr. Wheeler's paper, in which he gives the rainfall of 38 stations in that part of the country for twelve years—1869 to 1880 inclusive. Thus, if one-third be added to 26·05in., the maximum would be 34·73in., and if one-third be deducted the minimum would be 17·37in., which are almost exactly according to the rule; but then that would be expected, seeing that the rule itself is derived from observations of a similar kind, including these.

The elevation of the country at the source of these rivers is about 300ft. only above the sea. The geological formation in which these rivers rise, chiefly in the form of springs, is oolite, and they run over the Kimmeridge, Oxford, and Lias clays, and the glacial drift.

The fens, comprising 668,241 acres, are alluvium and peat. Vermuyden, a Dutch engineer, reclaimed a portion of this fen land, on a plan similar to plans adopted in Holland; but he made a great mistake in not taking into consideration the greater range of tide on this coast than on that of his own country, for as the tide ebbs out to a lower level here than it does there, he might have taken advantage of this to discharge the drainage direct into the estuary; instead of which, he merely raised embankments along the main rivers, and made long arterial cuts through the land to be reclaimed, placing at the end of each a sluice to keep out the tidal waters, each district or "level" dealing with its own drainage, irrespective of others alongside it, whereby conflicting interests were created, which have since caused enormous sums to be spent in litigation, have prevented a common action for the improvement of the rivers, and have increased the difficulties of river conservancy at the present day.

Another way of providing for the drainage of the fen lands was to abandon the winding course of a natural river, and make a long straight cut, shortening the distance, and increasing the rate of fall and the carrying capacity of the river. In these new rivers the flood-banks were set wide apart, the river channel occupying a comparatively narrow space in the centre, sufficient only for the ordinary discharge of water,|the intervening land between the river and the flood-banks being the "wash lands," over which floods spread and form shallow lakes of great extent, the water remaining on them for several weeks, in great floods, destroying crops and rendering the land of little value. When this wash land begins to dry the miasma arising from it is prejudicial to health, and it has become a nuisance.

The works necessary for the prevention of floods in these rivers require to be commenced at the outfall and carried on upwards through the whole length of the channel. The tidal waters enter freely three of these rivers, viz., the Ouse, the Nene, and the Welland, but in the Witham it is stopped at Boston by the "Grand Sluice," 8 miles from the sea.

The Ouse has a tidal course of 40 miles, nearly to Earith; in the Nene the tidal water flows 34 miles, to Northey Gravel, within $2\frac{3}{4}$ miles of Peterborough.

The Welland has a tidal course of 20 miles, as far as Crowland—all these distances, however, being those to which the highest spring tides reach. The sluice at Boston has self-acting doors, which open to let out the land drainage water, but close against the tide. The open navigation below the sluice has a depth of 16ft. at spring tides, but by the works carried out under the Witham Outfall Act of 1880 it is expected to give a navigable depth of 22ft. at the proposed entrance to the new docks at Boston.* For 20 miles above the sluice, to Bardney,

* This has since been effected.

the water is maintained for purposes of navigation at a uniform depth of 9ft.

The Grand Sluice had four openings of 16ft. each,* the depth of water on the sill at ordinary floods being about 10ft., rising to 14ft. in extreme floods. From Bardney to Boston the fall of the surface of the water in floods is from 3in. to 5in. per mile, and between Boston and the sea 25in. per mile. The works which have recently been carried out for making a new outfall by a cut $2\frac{1}{2}$ miles in length, and shortening the distance to the sea by $1\frac{1}{2}$ miles, were expected to lower the level of low water at the ends of the drainage cuts 3ft. at the least, by which the outfall sluices would continue open longer at every tide. The actual lowering of the water level has been more than 3ft. Lynn and Boston were once prominent ports, ranking only second to London and Bristol, and up to the time of the construction of railways there was a large export trade of wheat and agricultural products, and an import of coals and other goods, which were distributed throughout the midland part of England by these rivers; and Bedford by the Ouse, Northampton by the Nene, Stamford by the Welland, and Lincoln by the Witham, were placed in communication with the sea.

Now, here lies one of the difficult points of the whole question. So long as these navigations were maintained in order, and the shoals were cleaned out as they accumulated, and the locks and staunches preserved in efficient condition, and the weeds cut or kept down by the traffic of the boats, the rivers were capable of discharging the flood-waters, even in their artificial state of canalisation; but since railways have diverted the traffic from these inland rivers, navigation has ceased, the works have gone to ruin for want of funds to maintain them, and shoals and weeds choke the channels.

The rivers have become in a far worse condition to

---

* Another opening has been made in the "Grand Sluice" as part of the improvement works recently carried out under the directions of Mr. John Evelyn Williams, C.E.

discharge the drainage of the country than when left in their natural state, and constant floods are the consequence; but as there is a remnant of traffic left, the proprietors adhere to their rights in holding up the water, without having the means to adapt the rivers to the modern requirement of drainage by enlarging the capacity of the weirs, so as in times of flood to discharge waters which are now sent down into their district at a much greater rate than formerly.

This question opens into the arena of politics, and we do not pursue it further; but, like all such questions, it is an important one.

Mr. Jacob deals with a very different kind of river—the Irwell—which rises in the Erewell spring, near Bacup, in Lancashire, 1,325ft. above the mean level of the sea, and in its course towards the west coast receives the Limey water, the Roch, and the Croal, which passes through the town of Bolton, and after its junction with this stream the Irwell changes its direction, and runs a south-easterly course towards Salford and Manchester.

Comparing the inclination of the river with that of the surface of the watershed area through which it flows, Mr. Jacob says that the inclination of the river being the most gradual natural fall within this area, it has, of course, a less declivity than the average surface inclination of the ground. The mean inclination of the surface of the country at the source of the river is 540ft. to a mile; a few miles lower down the mean surface inclination is 380ft. to a mile, and on approaching Salford—that is, at Agecroft—where flooding usually begins, the mean surface inclination of the country is 340ft. to a mile. These represent the slopes at which surface-water descends into the brooks which feed the river. The inclination of the river itself at Bacup is 176ft. to a mile; between Bacup and Bury, 36ft., and immediately above Agecroft, 8ft. to a mile.

As the stream enters the borough of Salford the valley widens, and the inclination of the river bed is reduced to

4ft. to a mile. With these great inclinations in the upper parts of the river no great harm arose there from the cinders and other solid refuse got rid of by depositing it in the river or laying it on the bank within flood-level to be washed away; but the injury to the river lower down was much greater. The refuse which comes down the Irwell lodges in the flatter parts of the stream, and by raising the bed of the river throughout its course through Manchester and Salford has increased the injurious effect of floods.

There are certain weirs on the river to which it is usual to attribute the flooding; but their effect, Mr. Jacob says, upon the height of the floods has been much exaggerated. The removal of weirs is a desirable object, but as a rule there is no necessity, whilst improving a river, to deprive riparian owners of such water rights as they may have fairly acquired by time and right of user. Side sluices can without difficulty be constructed, or the weirs can be provided with opening bays. By such means interference with vested interests may, to some extent, be avoided, and those exorbitant claims set aside which so frequently compel administrative bodies to abandon works of improvement of the utmost importance. "The rights of private persons in respect of water and of easement are well defined in all ordinary cases. If any owner interferes with a stream, either by adding to or taking from its natural volume, so as to inflict injury upon his neighbours, there is no difficulty in compelling him to make reparation for the damage done. The fixed principle is that the stream may be lawfully used so long as others who enjoy the use of it are not prejudicially affected. It is only necessary to enlarge this principle sufficiently to show that almost every owner of property within the watershed of a river has a certain responsibility cast upon him in respect of that river," and this is in effect what we have said from the beginning in our remarks upon this subject.

On another point, also, what we have said agrees with

what Mr. Jacob and Mr. Wheeler said in these papers, and indeed, all those who took part in the discussion— twenty or more—viz., that whatever is done in improving the regimen of a river, it must be considered as a whole, and all the interests within the same watershed area taken into account. Upon the question of the storage of flood-waters in the upper and upper middle parts of river-basins, the circumstances of these particular rivers seem to warrant what was said about impounding flood-waters in those localities, but in this respect the circumstances of these rivers are exceptional.

Neither Mr. Wheeler's four rivers, nor Mr. Jacob's one river, afford proper sites for storage reservoirs, probably; but there are many others where proper sites do exist, and yet Mr. Wheeler said on this point:—"The existence of these wash lands, the large area they cover, and the above [those he had mentioned] facts are sufficient answers to those theorists who are in the habit of advocating the formation of reservoirs to regulate the streams and prevent floods. Here, on rivers draining a comparatively flat country, are occasional reservoirs of 3,000 and 5,000 acres, which yet have scarcely any effect in preventing most severe floods on the lands above them." That, no doubt, is so. But inasmuch as these so-called reservoirs are below the lands to be protected from floods, neither theorists nor practical men would expect any other effect from them. But the case is different where reservoirs are made above the lands intended to be protected by them.

And Mr. Jacob, on the same subject, said:—"It might at first appear easy to embank the valley above the flooded district so as to receive all the water in excess of what the channel is capable of discharging; but a very simple computation shows that the plan of impounding is unsuitable in such a watershed as that of the Irwell and its tributaries. A flood of ordinary duration lasts from ten to twelve hours at its highest level; and as there would be an excess in great floods of at least 8,000 cubic feet per second above what the channel is capable of discharging, it would be necessary to provide a reservoir to hold this quantity. . . . Mills and works of all kinds on the margin of the river would have to be purchased, railways, roads, and canals to be diverted, and altogether the outlay would be so formidable as to render the construction of reservoirs in the valley of the Irwell out of the question." Certainly, that would be so.

Such a thing, however, would not be attempted or thought of in the midst of mills, railways, and canals, and nothing, we believe, is intended to be conveyed by the "theorists" mentioned, more than that where suitable reservoir sites exist the flood-waters now flowing from the areas above them, to the detriment of the rivers, may be so regulated that the owners of those lands may substitute works for money in respect of their liability to taxation in contributing to the funds required for the conservancy of the rivers below them.

SECTION XXXI.

COUNTY BOARDS AND WATERSHED AREAS.

THROUGH the collateral remarks we have made on this subject, the main drift has been to consider how the prevention of floods may be effected, while at the same time making use, for domestic and industrial purposes, of the water held back from the rivers in such manner as to relieve them in flood-time of that excessive quantity which now pours into them, and which they cannot carry off without overflowing.

Of the causes of this overflow there are different opinions: some attribute it to the increased extent and thoroughness of land drainage; others to the obstructions of weirs or to narrow waterways of bridges, and withal to a general neglect of the watercourse itself, the banks of which have from time to time slipped into the river for want of protection, and the accumulated *débris* has been allowed to remain there, shifted in position from time to time, but never removed, the floods at one time clearing a passage here, while there the accumulation has been increased, the general effect being that the regimen of the river has been destroyed or deteriorated. And one of the causes of this deterioration has undoubtedly been the great quantity of manufacturing refuse which has been tipped into the river as the readiest way of getting rid of it; and house-ashes have been to a large extent got rid of in the same way.

Whether we look at the agricultural interest, the mining and manufacturing interests, the tenant or the landowner, millers, highway boards, county magistrates, railway companies, carriers by water or municipal authorities, we find them all implicated in the deterioration of the regimen of rivers, as well as the owners of the lands from which the water flows without regulation: and it is not one person, party, or body who should be called upon to repair and maintain the regimen of rivers, but all in the same river-basin, whether dwellers or receivers of rent, cultivators or owners of the soil; for all who have access to the river-banks contribute to its deterioration, while none does anything to improve it, or even to keep it in decent repair.

The river is the first and chief cause of prosperity of the dwellers in the valley, whether they directly use it or not, and even though some of them never see it. Its repair and maintenance, then, is the duty of all, and if each would make good the damage he does, whether actively or by permission only, no one would have any right to call upon him to do more; but, as a practicable thing, this is out of the question, and he must contribute to a common fund according to his means and liabilities, if he wishes to continue to enjoy his own rights in the river and what it brings him, without injuring his neighbour thereby. A good many people have not only wished to have all these advantages without doing anything in return, but have actually had them and converted them into hard cash. They should be allowed an opportunity now of contributing something to the repair of the damage they have done. The apportionment amongst all parties, so as to be just to all, is not in this case a matter upon which hairs can be split, but a near approximation to strict justice may be made without much trouble—without, that is to say, more trouble than the redress of a long-standing and general grievance must necessarily entail.

The authorities will find a difficulty in dealing with this question unless they form distinct watershed areas for rating purposes. This, however, they can do by an interchange of equal areas, or of areas somewhat approximately equal, which lie on the borders and upon both sides of the watershed line; not substantially, but fictitiously (have we no fictions in the Constitution already? Chiltern Hundreds to wit); not altering the existing boundaries of counties, where those lie on or near the watershed lines, but giving and receiving from each other such sums of money per annum as may be due from the small part of each county thus cut off, as it were, by the watershed line, or in some other way, for there are several ways of meeting this difficulty.

The boundaries of some of the counties in many parts follow the watershed lines, and so far there is no difficulty; but where these run through parts of counties, and re-enter adjoining ones, the principle of the majority might be acted upon—either a majority of votes or the greater part of the rateable value of property within these outlying parts. The inhabitants, if any, of such parts, might even have the choice to which county they will contribute, for the amount would be comparatively small, and the difference in the rates of one county and the adjoining one not great, perhaps, in any instance.

A greater difficulty than lies here will be found in those counties which have river boundaries. Where a river divides counties, the boundary is assumed to be the centre of its old course, in general, however its present course may differ from its old one; but neither in this case, nor in those where there has been no alteration in the position of the river, is this a well-marked line, nor indeed marked at all where the course of the river has not been wholly changed, and such a boundary may, under some circumstances, give rise to disputes between adjoining county boards. All dredging operations would meet with difficulties arising from the jealousy and distrust between one board and the other, or at least might do so, and as dredging and similar operations will, in many cases, be the first things to be done, it will be well not to open the way to this at the outset.

But a more frequent cause of difference of opinion will

be where any work of protection may be made on one bank of the river, and not simultaneously on the other bank in that reach of the river, for the action of water in rivers is such that, if one point is defended against its attack, it exerts its power upon another, and the easiest, which in such a case would probably be on the opposite bank a little lower down the stream, where the general line of the river is straight; while if both banks are defended its power is exerted upon the bottom, which is what is wanted, unless it be in exceptional cases.

There is, perhaps, no such thing as a long, old, straight, natural watercourse. If it is straight it is short. If long, it may be old, but is not straight. The ground is soft in some parts and hard in others, within the distance of only a few yards of a river-course, and the earth washed away from one part is deposited in another part of the river, and this usually takes place in the slack water it meets with at the bend of the river, the *débris* being deposited on the inner side of the bend, on whichever bank of the river that may be; but, as the general line of a river-course is straight, or nearly so, the deposit takes place at nearly the same number of places on each side of the river in the long-run. Were the ground on both sides of a river homogeneous, the course would not even then remain straight for any long distance, for, although it might remain so as long as neither bank were touched, and no projection from either were made in the slightest degree, yet, as soon as ever a small projection is made from either bank, the direction of the current begins to change, and, once begun, its natural action is to increase the inequalities.

Art, scientifically directed, can bring it back to its pristine form, or to such other form as wisdom derived from the experience of ages finds preferable; and that is what conservancy boards must now attempt.

Where the boundary is and must remain in the centre of the river, it may probably be found advisable to give control of both banks to one authority, or half the whole length to each adjoining one. A post on each bank, directly opposite each other, distinctly marks the division of jurisdiction, while the centre of the river is but a vague and uncertain division; and even if the two adjoining authorities could always agree to carry on the work of protection or repair of the banks so that no injury should be done by one to the other side, there are other, though less important, difficulties which would arise out of such a boundary, irrespective of any question concerning the banks.

It cannot be contemplated that any conservancy board would propose to alter county boundaries so far as to remove all river boundaries to their cognate watersheds; otherwise many advantages would result to the management of rivers and all streams, by making new divisions between such counties coincident with the watershed lines, which are capable of being accurately defined and permanently marked. Our control of water is limited; we are masters of it only after it has fallen; we turn it this way and that, but we cannot in any important degree change the ridges of the hills, and must take from the natural watershed what comes.

Having got it, we can carry it from one valley to another, or from one side of a valley to the opposite one, and in doing this the work would be much facilitated if one board had the command of the whole of one valley. In preventing floods, such a jurisdiction would be equally advantageous; indeed, the advantages in this respect would be greater than in the other, for while a board on one side of a river may do all that is necessary for this purpose, and do it economically, they cannot economically carry out effectual works without the co-operation of the board on the other side of the river.

Inasmuch as the abstraction of water from rivers or streams for the supply of towns, or indeed for any purpose, diminishes the quantity of flood-water in rivers, waterworks ought to be exempt from contribution to funds raised for the purpose of preventing floods. Where waterworks are situated on the banks of rivers, the water is not taken in flood-times, for it is then muddy, and the sluices are closed; and in any case the quantity abstracted would be so small, relatively to the volume of the flood, that the effect would be of no importance. But in gravitation works the case is different; here the flood-waters are impounded as a necessary part of the operation of the works, and if the storage reservoirs are sufficiently capacious to hold the great floods which at present cause so much injury, they assist the operations of any floods prevention Act which may be passed. But of course a large quantity of water may be abstracted without diminishing injurious floods at all. A reservoir which has a capacity sufficient to hold the floods which occur during any three consecutive years of least rainfall only, is not large enough to retain the floods of wet years, or even of average years, for by the very terms upon which its size has been calculated, a portion of the average rainfall is allowed to pass it in floods which are omitted from the calculation upon which the capacity of the reservoir has been based.

It used to be that the average rainfall was intended to be impounded, except that part which is evaporated and otherwise lost before the water arrives at the site of the reservoir, and this of course included all floods, great and small. But the inequalities of rainfall of different years in a long series are so great, that to make reservoirs large enough to effect this, they would have cost too much money; and the system was adopted of calculating upon the regulation of the flow of water proceeding from the average rainfall of three years only, these being the driest three years to be found together in the records of the rainfall. It came to be known by experience that this was so nearly five-sixths of the average of the whole series, that a system was adopted of deducting at once a sixth part of the average annual rainfall, as being the quantity which could be stored in reservoirs of any practicable dimensions—practicable, that is to say, with such an outlay of money as could be afforded. This, no doubt, was a very proper thing to do; but it is inconsistent with a claim to exemption, or at least to total exemption, for, admittedly, the floods of wet years are not stored, to the extent of one-sixth of the whole rainfall, less the amount of evaporation and absorption, and this one-sixth part is a very large part in volume; and it occurs, moreover, at those times when the benefit of its being stored or regulated would be greatest, and when its liberation does most harm.

To take an instance, where the average annual rainfall of a locality is 48in., it might be 40in. only upon which the capacity of a storage reservoir would be based, after deducting the proper depth for evaporation and absorption; but as to this latter, it only means the absorption of so much as does not reappear within the drainage area, and is a small quantity in any case. The two together, and that which is taken up by vegetation-growth, are better called the loss; and this might be in such a case reckoned to be 13in., leaving 27in. only with which the capacity of the reservoir would be concerned, a portion of this being abstracted daily for the main purpose and diverted from the stream, and another portion left daily to the stream, its delivery out of the reservoir in the same quantity during dry weather, as at other times, forming the compensation to the interests in the stream for the abstraction of the other portion, say, two-thirds abstracted and one-third left, or 18in. abstracted and 9in. left to the stream during three consecutive years of least rainfall. There would thus be 8in. in depth over the drainage area, which would pass the reservoir in floods, in a year of average rainfall, and the average annual quantity per 1,000 acres of the drainage area would thus be 29,000,000 cubic feet, which cannot be stored by reservoirs of these dimensions.

But if the average annual rainfall be 48in., and the mean of any three years be 40in., there must be other years in which the rainfall exceeds 48in., and examinations of rainfall records show that the depth of rain falling in the wettest year is about twice as much as in the driest year of all. In the above case 65in. or 66in. might be the depth fallen in the wettest year, and 33in. in the driest, the mean of all years being 48in., while the actual quantity dealt with by the construction of a reservoir would be about 40in., less the amount of evaporation, &c. If we do not take into consideration the very wettest year of all in the record, but a depth which occurs rather frequently, viz., 60in., there would be in each of those years, however many or few, 20in. of rainfall unimpounded, if the same loss by evaporation, &c., be allowed in these years as in the average of years, and, per 1,000 acres, the quantity is 72,600,000 cubic feet of water, which cannot be stored in reservoirs, the capacities whereof have been calculated on those bases, and this water must continue to go down the river in floods. Hence, a total exemption of such waterworks cannot be claimed. But it may be said that the effect of great floods is at least mitigated by such reservoirs, and that is true, for even when the reservoir is full, a flood does not at once issue from it at the same rate as that with which it comes into it; it expands over the area of the reservoir and raises its surface, and the time taken up in doing so is additional to the time in which the same volume of water would otherwise pass down the river, thus reducing its rate of flow.

The areas of water surface of reservoirs having different areas of watershed do not vary in the same ratio as the watershed areas, and therefore no simple number of acres of water surface per 1,000 acres of watershed area can be stated for all circumstances; but we may take an instance in which there are 40 acres of water surface per 1,000 acres of the watershed area, and a flood running into the reservoir at the rate of 360,000 cubic feet per hour, or 6,000 cubic feet per minute, for 24 hours. A much greater quantity per minute might pass during some portion of the time. Supposing the reservoir full at the commencement of the flood, the water-level would be raised 18in. in from 7 to 8 hours, if none passed out during the time. If the waste-weir of the reservoir for this drainage area were 20ft. wide (and it would not be safe to make it less), and if the co-efficient of discharge over it be taken at ·5, which is the proper co-efficient for a weir with a crest 3ft. wide, sloping 1 in 18 across, the above-named quantity of 360,000 cubic feet per hour would pass over at a depth of 18in., and the rate of flow out of the reservoir would be the same as that into it, when the water reached that height. Considering the quantity which would be passing out of the reservoir while 360,000 cubic feet per hour would be coming in, the time during which it would continue to rise would be extended to about 12 hours, after which it would run out as fast as it ran in, which would be during the next 12 hours, and on the cessation of the flood the accumulated water would gradually subside to the level of the waste-weir. Thus in this case the claim for abatement of the effect of this flood would be in respect of half the time only.

The case of the new Vyrnwy Waterworks for Liverpool is apropos. Mr. Sargeaunt, of Tewkesbury Park, a magistrate for the county of Gloucester and of the borough of Tewkesbury, in his evidence before the Committee on the Liverpool Corporation Water Bill, in 1880, said that the damage by floods to meadow-land on the banks of the Severn between Gloucester and Worcester had been half a million of money, between the years 1871 and 1879, and he supported the Bill because he thought it would diminish floods. The Severn Commissioners, like most other people in the valley, when notice was given of the intention of the Liverpool Corporation to abstract water from the head of the river, were apprehensive of serious losses which might result to them, especially in dry weather; but also in the case of the Navigation Commissioners, in the effect of the works in diminishing the scour of the bed of the river; and they consulted Mr. (now Sir) John Fowler, C.E., who reported to them that the effect which the proposed works would have upon the Severn between Stourport and Gloucester would be a slightly-increased flow at low water, and a diminution, to a certain extent, in the frequency and height of the floods. He estimated, from the data then available to him, that a "bank-full" flood (16ft. upon the upper sill of the Diglis Lock) would be diminished about 6in., and a "high flood" (20ft. upon the sill) about 5in.; and he was of opinion, therefore, that the navigation interests would be practically unaffected, either for good or evil, by the proposed works. The quantity of water at that time proposed to be abstracted was 52,000,000 gallons per day. This was considerably reduced on the demands of those interested in the river, and the amount of compensation-water left to the stream correspondingly increased.

On the basis of the probable annual discharge of the river at Diglis in 1868, an exceptionally dry year, and in 1872, an exceptionally wet year, and on the measurements of Mr. E. Leader Williams, C.E., of the discharge at the same place at low water, bank-full, and high-flood, Sir John Fowler estimated the mean discharge of the Severn at Diglis to be 120,000 cubic feet per minute in an exceptionally dry year; 290,000 in an exceptionally wet year; and the average of a series of years to be 164,000 cubic feet per minute. At Gloucester, although the drainage area is nearly twice as large as at Diglis, near Worcester, he estimated the discharge to be not more than 168,000 in a dry year; 464,000 in a wet year; and 246,000 cubic feet per minute on an average of a series of years, being respectively 40 per cent., 60 per cent., and 50 per cent. greater than the quantities at Diglis; and having thus obtained approximate estimates of the mean discharge of the river, it became possible to him to ascertain the extent of the interference incidental to the proposed Vyrnwy works of the Liverpool Corporation.

Assuming that 52,000,000 gallons a day would be abstracted, that would be 5,800 cubic feet per minute, or 4·83 per cent. of the whole quantity going over Diglis Weir in dry years, 2 per cent. in wet years, and 3·54 per cent. in average years, whilst the drainage area affected by the proposed works was only 1·84 per cent.

AN EXCEPTIONALLY LARGE RESERVOIR.

In like manner the flow at Gloucester would be affected to the extent of 2·45 per cent. in dry years, 1·25 per cent. in wet years, and 2·35 per cent. in average years, the drainage area affected being 6·92 per cent. of the entire river-basin.

With respect to floods, Sir John Fowler was of opinion that 1½in. depth of rainfall in 24 hours running off the Vyrnwy watershed area at the site of the proposed reservoir (Llanwddyn) would probably be a full measure of the maximum flood discharge in any year, which would amount to less than 90,000 cubic feet per minute, being about five times the low-water discharge over Diglis Weir; but comparing it with Mr. Leader Williams's gaugings it amounts to but 13 per cent. of the bank-full flood, and 7·8 per cent. of the high flood at Diglis, and, of course, a correspondingly smaller percentage of the flood discharge at Gloucester.

It was stated by Mr. Hawksley, that it is intended to take for Liverpool an average daily quantity of 36 million gallons; about 40 millions in summer, and 32 millions in winter, giving to the stream 13½ million gallons a day on the average of each year. It was originally intended only to give 8 millions to the stream, but this has been increased to 10 millions every day, and 32 flushings in the year, each of 40 million gallons. The command of water is very great in these works.

The drainage area of the Vyrnwy alone is 17,583 acres at the site of the reservoir, and the average annual rainfall 61in.; but besides this large area, powers were taken by the Act to add 2,880 acres of the watershed area of the Afon Conwy, and 1,537 acres of that of the Marchnant, by driving a tunnel through the separating hills in each direction, so as to bring the water into the Llanwddyn Reservoir, which was to have a water surface of 1,115 acres.

Sir F. J. Bramwell, C.E., who supported Mr. Hawksley's evidence, said of this reservoir that in times of drought it will increase the flow of water in the river probably five times; that it will frequently prevent floods, as far as they arise in the watershed area belonging to it, from arising at all, and in wet years, when they do occur, they will never reach their present intensity.

But he said also, which is important to observe, that the bulk of the water arising from the average annual rainfall of 61in. comes down in floods. For instance, 31 per cent. of all the water that comes down, comes when the river is discharging at the rate of 100,000,000 gallons a day, or over, and this 31 per cent. of the quantity of water comes down in 18 per cent. of the time.

Since the evidence upon which these remarks are founded was given, the large works to which it refers have been carried out towards completion, under the directions of Mr. Deacon, the Waterworks Engineer of Liverpool, and some alteration of the quantities of water to be taken and left, respectively, appear to have been made; but the original quantities stated serve well enough for comparison and for a general statement of the case, which is all that is here intended.

The chief object of this reservoir is the water-supply of Liverpool, and it is not constructed in any respect with the object of preventing floods in the valley below it; but its relative area being so large, viz., about 50 acres per 1000 acres of its watershed area, it will have the effect, in cases of heavy rainstorms within its watershed area, of mitigating floods below it almost as much as if special provision had been made for that purpose. But its circumstances are very exceptional; it is nearly five miles long, and is not here adduced as having any bearing on the subject of flood-regulation elsewhere.

INDEX.

AIR-LOCKS in conduit pipes 88, 89
Aqueducts, or large conduits, 113–119
Areas of river-basins, 72–77

BARROWS, 10
Bed-puddle, 94
Blackwell's experiments, 53, 59
Bottom velocity, 125, 126
Bradfield Reservoir, 13, 30
Branch mains, 103
Breast wheels, 156, 158, 175, 179–183
Buckets of wheels, 140
Bye-wash, 43

CAPACITIES of reservoirs, 24, 283
Carts, 10
Cast-iron pipes, 104–112
Clay puddle, 2, 3
Co-efficients of discharge, 59–61

Concrete, 47–50
Conduits, 79–87
—— large, or aqueducts, 113–119
Conical tube, 150
Conservancy of rivers, 287–294
Consolidation of embankments, 10–12
Contraction of jet, 146–148
Cornish engine, 244
Cost of a reservoir (approximate), 44, 267–269
Current wheels, 160–162, 166, 167

DAYS' storage (number of), 18, 20
Depth of water on the outer edge of a weir, 62
Detection of waste of water, 237
Discharge pipes, 5, 31, 32, 40, 41
Dobbin carts, 10
Domestic water-supply, 217–221
Du Buat's formula, 121, 123

EFFECTIVE fall on wheel, 142
    Efflux of water from a hole, 146
Embankments, 8–13
Excavating puddle-trench, 35
Eytelwein's formula, 101

FLOODS, 247–286
    Flow of water off the ground, 253–258
Freezing of open channels, 82

GAUGES (rain), 70, 71
  —— (water), 51–63
Gravel puddle, 3, 4
Grimwith Compensation Reservoir, 29
Grindstones driven by water-power, 186–191

HEAD of water, 99
    High-breast wheels, 141, 177–179
High falls of water, 192
Holmfirth Reservoir, 13
Horse-power, basis of, 136
Hydraulic gradient, 102
—— mean depth, 81
Hydraulicians (the older), 147

INTERSTICES of concrete materials, 48–50
Irwell, river, 291

KUTTER'S formula, 122

LAND, relief of, from floods, 271–277
Llanwyddyn Reservoir, 305.

METERS, 234–236
    Mills, 127–216
Motion of water in pipes, 100

NEVILLE's formula, 122, 125

PIPE across puddle trench, 33
    Pit sand, 46
Plank gauges, 51–58
Poncelet's wheel, 154
Pressure in pipes, 106–109
—— of the atmosphere, 97, 98
Prony's formula, 124
Puddle lining, 227
—— trench, 4, 36, 37
—— wall, 1
Puddled clay, 2, 3
Pumping main, 241
—— water by water-power, 184, 185

RAIL waggons, 10
    Rainfall, 17, 64–69
Rain gauges, 70. 71

INDEX. 309

Regulation of flood waters, 278-286
Relief of land from floods, 271-277
River *in train*, 137, 163, 164
River sand, 46
—— Irwell, 291
—— Severn, 259
—— Thames, 259
—— Witham, 289
Rivers as county boundaries, 298
—— conservancy, 287-294
——, lowering the water-level of, 248
——, yearly neglect of, 248
Road watering, 233
Rotative beam-engine, 245

SEPARATING the clear water, 128
Service reservoirs, 93-95, 222-229
Severn and Thames contrasted, 260
Slip-joints, 40
Standpipes, 243
Storage per acre of gathering ground, 21
Strainers, 95
Stream gauges, 51-63
Strength of cast-iron pipes, 104-112
Surface velocity, 124

THAMES and Severn compared, 259

Thickness of cast-iron pipes, 110-112
Tittesworth Reservoir, 30
Top-bank above top-water, 8, 23
Trials of coal consumed in pumping water, 245, 246
—— of water required to work mills, 171, 174
Tunnels, 90, 91
Turbines, 192-216

UNDERSHOT water-wheels, 152, 167, 170, 174

VALVE pit, 30
—— tower, 42
Velocity in open channels at the surface, 120-127
—— in open channels, the mean, 120-126
—— in open channels at the bottom, 120-126
—— of approach to weirs, 158
—— of water, 165
—— of water in pipes, 99-101

WASTE of water, 237
Watering roads, 233
Weight of water, 98, 135
Weirs of reservoirs, 28-30
—— on rivers, 251
Witham, river, 289

www.ingramcontent.com/pod-product-compliance
Lightning Source LLC
Chambersburg PA
CBHW030806230426
43667CB00008B/1092